Predicting Vehicle Trajectory

Predicting Vehicle Trajectory

Cesar Barrios
Yuichi Motai

CRC Press
Taylor & Francis Group
Boca Raton London New York

CRC Press is an imprint of the
Taylor & Francis Group, an **informa** business

CRC Press
Taylor & Francis Group
6000 Broken Sound Parkway NW, Suite 300
Boca Raton, FL 33487-2742

First issued in paperback 2020

ISBN 13: 978-0-367-65634-8 (pbk)
ISBN 13: 978-1-138-03019-0 (hbk)

Library of Congress Cataloging–in–Publication Data

Names: Barrios, Cesar, author. | Motai, Yuichi, author.
Title: Predicting vehicle trajectory / Cesar Barrios and Yuichi Motai.
Description: Boca Raton : Taylor & Francis, a CRC title, part of the Taylor & Francis imprint, a member of the Taylor & Francis Group, the academic division of T&F Informa, plc [2017] | Includes bibliographical references and index.
Identifiers: LCCN 2016043395| ISBN 9781138030190 (hardback : acid-free paper) | ISBN 9781138031623 (ebook)
Subjects: LCSH: Automobiles--Collision avoidance systems. | Trajectories (Mechanics)--Data processing. | Automotive computers.
Classification: LCC TL272.52 .M68 2017 | DDC 629.2/042--dc23
LC record available at https://lccn.loc.gov/2016043395

Visit the Taylor & Francis Web site at
http://www.taylorandfrancis.com

and the CRC Press Web site at
http://www.crcpress.com

Prediction is very difficult, especially if it's about the future.

—Niels Bohr, Nobel laureate in physics

This book is dedicated to the love of my life, Raquel,

for her unconditional encouragement.

—Cesar Barrios

I dedicate this book to my wife for her unlimited support.

—Yuichi Motai

Contents

Preface

Prediction of the trajectory path of a vehicle into the future is a difficult task, and even more so in non-straight paths, as observed in some of the research studied. Many times, the predicted future position of where the vehicle will be 3 s later in time falls outside of a physical road, making this prediction highly improbable. For the first part of this research, the assumption is made that the driven vehicle will remain on a road at all times, and any prediction that falls outside of a road will be considered incorrect. Through the use of a road-mapping technique, it will be shown that this error correction greatly reduces the prediction errors in non-straight paths.

Another problem observed when predicting a future position of a vehicle is that, when using multiple sensors, most of the time they are asynchronous. Some reviewed research describes a solution of running the system at the rate of its slowest sensor, and, therefore, solving the problem of asynchronous data. Other reviewed research uses previously estimated measurements to fill in the missing data from offline sensors. A vehicle is a large object that cannot change its spatial dynamics very quickly, but running a prediction system at a slow rate can slow down the detection of these spatial changes. For this research, the system is run at the rate of its fastest sensor, but missing measurements are calculated based on measurements obtained from online sensors using a dead-reckoning approach. A technique was developed to properly handle error accumulation from missing data from offline sensors, and that running the system at the fastest rate possible greatly reduces the prediction errors in non-straight paths.

The last part of this research looks into a possible solution to advance the usability of a vehicle-to-vehicle (V2V) system in its initial stages. The National Highway Safety Administration announced in 2014, its decision to begin taking the next steps toward implementing V2V technology in all new cars and trucks. Although all vehicle manufacturers are required to support this technology, it will still take many years until the V2V system is fully populated and most vehicles can participate. Until that point is reached, the benefits of the V2V technology will not be taken advantage of, unless a temporary solution is achieved to enable older vehicles to participate in the V2V system as well. Smartphones are readily available and already have many built-in sensors and good processing power, so in this part of the research, the possibility of using smartphones to predict the trajectory path of a vehicle will be used. It will be shown that some kinds of smartphones yield similar prediction errors as predictions calculated using vehicle-mounted sensors.

MATLAB® is a registered trademarks of The MathWorks, Inc. For product information, please contact

The MathWorks, Inc.
3 Apple Hill Drive
Natick, MA 01760-2098
USA
Tel: 508-647-7000
Fax: 508-647-7001
E-mail: info@mathworks.com
Web: www.mathworks.com

Acknowledgments

Dr. Cesar Barrios' work was funded in part by the U.S. Department of Transportation through Lisa Aultman-Hall, director of the University of Vermont Transportation Research Center. Dr. Yuichi Motai received support from the National Science Foundation Grant #1054333.

The authors want to especially acknowledge Dr. Dryver Huston from the University of Vermont, and Dr. Adel Sadek from the University of Buffalo, for their invaluable comments on parts of this work and Dr. Walter Varhue, from the University of Vermont, for his contributions as the committee chairman.

The authors also acknowledge Dr. Eric Jackson for sharing some log files of data collected at the University of Connecticut, and Dr. Henry Himberg and Samuel Lopez for their comments on Kalman filters.

Authors

Cesar Barrios received his BS (1999) and MS (2001) degrees in electrical engineering from the New Jersey Institute of Technology and his PhD (2014) in electrical engineering from the University of Vermont. After receiving his BS degree in 1999, he worked for IBM. Dr. Barrios has worked for Global Foundries since 2015. He first started in information technology, and then moved into semiconductor research and development.

Yuichi Motai received his BEng in instrumentation engineering from Keio University, Tokyo, Japan in 1991, an MEng in applied systems science from Kyoto University, Kyoto, Japan in 1993, and a PhD in electrical and computer engineering from Purdue University, West Lafayette, Indiana, USA, in 2002. He is currently an associate professor of electrical and computer engineering at Virginia Commonwealth University, Richmond, Virginia, USA. His research interests include the broad area of sensory intelligence, particularly in medical imaging, pattern recognition, computer vision, and sensory-based robotics.

1

*Improving Estimation of Vehicle Trajectory Using the Latest Global Positioning System with Kalman Filtering**

1.1 Introduction

Accurately predicting the future location of a vehicle is a very important and relatively difficult process in the intelligent transportation system (ITS). It can be effectively used in obstacle avoidance systems for vehicles or robots.

Many of the existing obstacle avoidance systems currently being researched are limited to line-of-sight sensors, such as those described in References 1 through 9, using sensors around the vehicles to detect nearby objects. For a long-range obstacle avoidance system, other types of sensors need to be implemented such as those presented in References 10 and 11.

Research studies like those conducted at the Kansai University of Japan [10] or those by Miller and Oingfeng [11] investigates the option of using global positioning system (GPS) data collected from different vehicles to predict the future location of each vehicle. The methods used to make these predictions are somewhat simple and do not give very accurate results in scenarios such as curves, shown in Figures 1.1 and 1.2, where the estimated future position of the vehicles will not be a straight path.

It is clear from the current research that what is needed is a more accurate way to predict the trajectory of vehicles in all different scenarios. This is where the Kalman filter (KF) comes into play. The KF has a long history of accurately predicting future states of a moving object and has been applied to many different fields, which is why it has been chosen for this research [12–15].

The aim of this chapter is to investigate the viable idea of using geographic information system (GIS) to reduce error in the prediction of the future location of an automobile, particularly during curves. The system implemented

* This chapter is a revised version of the author's paper in IEEE Transactions on Instrumentation and Measurement. DOI: 10.1109/TIM.2011.2147670, approved by IEEE Intellectual Property Rights.

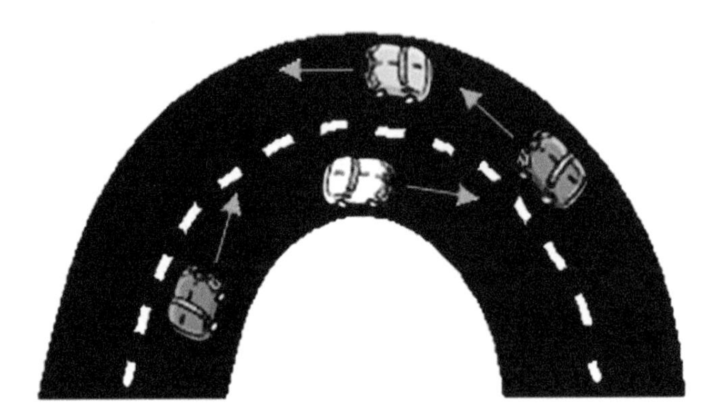

FIGURE 1.1
"C" crossing.

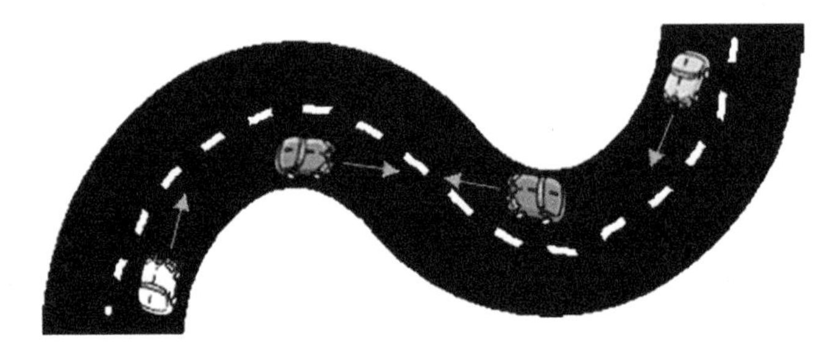

FIGURE 1.2
"S" crossing.

in this chapter consists of an interacting multiple model (IMM) estimation with different KFs using the GPS to get a vehicle's spatial information.

There are a number of existing studies concerning the best methods to take spatial coordinates that fall outside of a defined road and to estimate where they would fall on an actual road, also known as map-matching. For example, in References 16 through 19, the authors go into a lot of detail to explain the different errors that need to be accounted for when using a GPS sensor (among others) and data for road maps (GIS), and how the GPS bias can be utilized to improve the map-matching accuracy. Other researchers, such as References 20 and 21, look into the problem of GPS outages, and how the vehicle's position can be estimated during the outage through the use of KF and map-matching techniques. This study compares experimental results of predictions done with and without our GIS error correction algorithm, and does not consider the problem of GPS outages since other researchers are working solely on this issue.

The GPS is a satellite-based navigation system made up of a network of 24 satellites placed into orbit by the U.S. Department of Defense. These GPS satellites circle the earth twice a day in a very precise orbit and transmit signal information to the earth. The GPS receivers take this information and use triangulation to calculate the user's exact location. There are several factors that can degrade the GPS signal and, thus, affect accuracy, the most important being ionosphere and troposphere delays. These delays can contribute up to 10 m to the inaccuracy of the position, but some newer GPS receivers have some error correction systems to improve the accuracy.

This research is based on the use of a GPS receiver to obtain location information and to be able to estimate the projected path for a vehicle; so, selecting a good GPS receiver was of utmost importance. The Holux GR-213 1 Hz GPS receiver used in this research is wide area augmentation system (WAAS) enabled. The WAAS is a system developed for civil aviation by the Federal Aviation Administration (FAA) in conjunction with the U.S. Department of Transportation (USDOT). It is a nationwide differential GPS system where base stations with fixed receivers calculate and transmit the GPS error to the geostationary satellites in its view, which in turn broadcast the corrections that can be used by individual WAAS-capable GPS receivers. Its accuracy is less than 3 m 95% of the time, and the GPS receiver used claims to have an accuracy of less than 2.2 m [22].

Similar systems designed to predict a vehicle's trajectory implement the use of other types of sensors to be able to get an accurate estimation, but this research looks into the possibility of using a commercially available, inexpensive but accurate GPS receiver to do a similar task already implemented in some areas [12,14,15,23–25], and it takes advantage of using a location-based system, such as knowing where on a road map the vehicle is located.

To be able to predict a vehicle's future location, the KF was used. The KF is a set of mathematical equations that provides an efficient computational (recursive) method to estimate the future state of a process. The filter is very powerful because it supports estimations of past, present, and even future states, and it can do so even when the precise nature of the modeled system is unknown [25–34].

The multiple KF models approach was chosen over one complex model because setting up multiple smaller models for each different scenario would be simpler than defining one complex model that can be accurate in many different scenarios. Each simple model is good for one specific set of conditions, so several models need to be defined to be able to cover most, if not all, possible scenarios in which a vehicle can be found. For this setup, four models have been identified to cover most of the vehicles' behaviors: a vehicle not moving; a vehicle traveling at constant velocity; or with constant acceleration; or with constant jerk (constant change in acceleration). These models provide a mathematical set of equations that can be used to predict the vehicle's future location after a set amount of time (Δk).

This study researches trajectory estimation at 3 s ahead in time, based on the average 1.5 s human reaction time to be able to react and avoid an accident [35]. The 3 s ahead in time was chosen as a reference point that is double the reaction time of an average human being. In reality, this number will probably vary in relation to the speed and type of the vehicle, since a faster or heavier vehicle will need more time to slow down. The fastest data rate of the GPS receiver used is 1 s ($\Delta k = 1$), so that is the rate the system will run at, which is set up to estimate the location of the vehicle 1 s later in time. To be able to obtain an estimation for the location of a vehicle 3 s later in time, the researchers needed to run the prediction steps of the KF system with Δk set to 3 s, and use the IMM to obtain the prediction. This extra step to estimate the 3-s ahead location adds very little runtime to the overall system, since it is only used to predict the location and no correct steps are needed.

1.2 Kalman Filter

The KF estimates a process by using a form of feedback control loop: the filter estimates the process state at some time, and then obtains feedback in the form of (noisy) measurements, and then it repeats (see Figure 1.3). As such, the equations for the KF fall into two groups: what we have called "prediction step" and "correction step."

The prediction step equations are responsible for projecting forward (in time) the current state and error covariance estimates to obtain the *a priori*

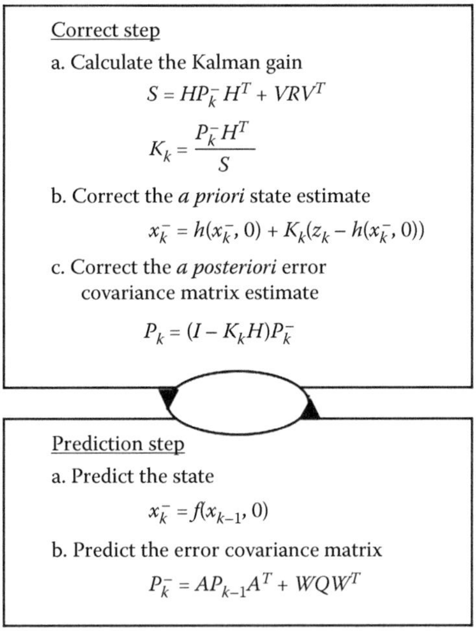

Correct step

a. Calculate the Kalman gain

$$S = HP_k^- H^T + VRV^T$$

$$K_k = \frac{P_k^- H^T}{S}$$

b. Correct the *a priori* state estimate

$$x_k^- = h(x_k^-, 0) + K_k(z_k - h(x_k^-, 0))$$

c. Correct the *a posteriori* error covariance matrix estimate

$$P_k = (I - K_k H)P_k^-$$

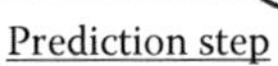

Prediction step

a. Predict the state

$$x_k^- = f(x_{k-1}, 0)$$

b. Predict the error covariance matrix

$$P_k^- = AP_{k-1}A^T + WQW^T$$

FIGURE 1.3
Extended Kalman filter.

estimates for the next time step. The correction step equations are responsible for the feedback—that is, for incorporating a new measurement into the *a priori* estimate to obtain an improved *a posteriori* estimate.

Correction Step (Using Measurement Data)

- Compute a gain factor (*Kalman gain*) that minimizes the error covariance.
- Correct state estimation by adding the product of the Kalman gain and the prediction error to the prediction.

Prediction Step (From the State Variables)

- Predict the next state from the current state using the system model. Assume the model is perfect (no process noise).
- Predict the error covariance of the next state prediction.
- Correct the error covariance estimation using the Kalman gain.

For our system, the state vector for this system consists of two parameters obtained from the GPS sensor, each one decomposed into its x and y components. The general form of the state estimate matrix is shown in Equation 1.1:

$$x = \begin{bmatrix} x_v \\ v_v \end{bmatrix} = \begin{bmatrix} \text{Position of vehicle} \\ \text{Velocity of vehicle} \end{bmatrix} \tag{1.1}$$

The elements of the state vector in Equation 1.1 were selected to account for all the measurements available from the GPS sensor, and from them any other variables needed for the KF models were derived. Keep in mind that each of the components of the state estimate in Equation 1.1 has an x and y component to it. So for every x_k represented in the equations, there will be an x_{kx} and an x_{ky}.

The error covariance matrix is a dataset that specifies the correlations in the observation errors between all possible pairs of vertical levels. The error covariance for each KF was approximated by running the filters on their own, but its value gets adjusted every 1 s in our setup.

The estimated error covariance P (1.2) is used together with the Jacobian matrix of the measurement model H (1.3 and 1.4) and the measurement noise covariance R (1.5) together with the Jacobian matrix of the measurement model with respect to the measurement noise V (1.6) to calculate the Kalman gain K:

$$P = \begin{bmatrix} x_v x_v & x_v v_v \\ v_v x_v & v_v v_v \end{bmatrix} \tag{1.2}$$

$$h(x,v)=\begin{bmatrix} x_v+v \\ v_v+v \end{bmatrix} \tag{1.3}$$

$$H=\left[\frac{\partial}{\partial x}h(x,0)\right]_{\substack{x=x(k-1)\\ v=0}} \tag{1.4}$$

$$R=\sigma_m^2\cdot\begin{bmatrix} I & 0 \\ 0 & I \end{bmatrix} \tag{1.5}$$

$$V=\left[\frac{\partial}{\partial v}h(x,0)\right]_{\substack{x=x(k-1)\\ v=0}} \tag{1.6}$$

Once the Kalman gain (K) is calculated, the system looks at the measured data z (1.7) to correct the predicted position and also the covariance error. Since this system only obtains measurements from a GPS receiver, only location, speed, and heading angle can be obtained; therefore, the other two parameters need to be calculated from the measured data. The acceleration is calculated from the velocity difference between the current and previous reading, and similarly, the jerk is calculated from the acceleration difference between the current and previous values. For this experiment, instead of using the current speed and heading from the GPS sensor, the average speed parameter was also similarly derived from the location difference between the current and previous values:

$$z=\begin{bmatrix} x_v \\ v_v \\ a_v \\ j_v \end{bmatrix} \tag{1.7}$$

Another important item to point out is that this research does not look into solving the GPS measurement errors that are due to many factors. One of these error contributors is the "signal multipath" problem, where the signal reflects off large objects. In this research, it is assumed that these errors are minimal since the experiment is done in a very rural area. Also, signal delays (ionosphere and troposphere) can cause the location readings from the GPS to bounce around and imply movement when the vehicle is not even moving. There are many error contributors to the GPS receivers, but we will assume them negligible in this research to concentrate on the main objective of this system.

After correction of the previously predicted values, the system is ready to predict the next position by using the state vector equations. The filter also estimates the error covariance of the estimated location by using the Jacobian matrix A (1.8), and the Jacobian matrix W (1.9), together with the process noise covariance Q (1.10).

$$A = \left[\frac{\partial}{\partial x} f(x,w) \right]_{\substack{x=x(k-1) \\ w=0}} \tag{1.8}$$

$$W = \left[\frac{\partial}{\partial w} f(x,w) \right]_{\substack{x=x(k-1) \\ w=0}} \tag{1.9}$$

$$Q = \sigma_p^2 \cdot I \tag{1.10}$$

To obtain an accurate prediction of the vehicle's future location, four adaptive prediction algorithms are defined to account for the possible scenarios. The state equations are very different between the models. The following four models account for most, if not all, possible situations in which a vehicle could be found.

Constant Location Model (CL)

$$\begin{aligned} x_v(k) &= x_v(k-1) + w \cdot \Delta k \\ v_v(k) &= 0 \end{aligned} \tag{1.11}$$

Constant Velocity Model (CV)

$$\begin{aligned} x_v(k) &= x_v(k-1) + (v_v(k-1) + w)\Delta k \\ v_v(k) &= v_v(k-1) + w \end{aligned} \tag{1.12}$$

Constant Acceleration Model (CA)

$$\begin{aligned} x_v(k) &= x_v(k-1) + v_v(k-1)\Delta k + \frac{1}{2}(a_v(k-1) + w)\Delta k^2 \\ v_v(k) &= v_v(k-1) + (a_v(k-1) + w)\Delta k \end{aligned} \tag{1.13}$$

Constant Jerk Model (CJ)

$$\begin{aligned} x_v(k) &= x_v(k-1) + v_v(k-1)\Delta k + \frac{1}{2} a_v(k-1)\Delta k^2 + \frac{1}{6}(j_v(k-1) + w)\Delta k^3 \\ v_v(k) &= v_v(k-1) + a_v(k-1)\Delta k + \frac{1}{2}(j(k-1) + w)\Delta k^2 \end{aligned} \tag{1.14}$$

In Equations 1.11 through 1.14, Δk represents the period of time passed, so the variables at $k - 1$ represent the data from 1 period of time ago. In this setup, the period of time is driven by the data rate of the GPS (1 s). The process noise covariance for each of the models (w) is based on the constant term only. For example, for the CV model, the process noise covariance is based on the velocity term only, and it can be derived from the measured data by applying the CV model to it.

Equations 1.11 through 1.14 represent the four states in which a vehicle can be found: at rest; moving at constant velocity; moving at constant acceleration; or moving at constant jerk. Each of these models consists of four state equations used to calculate each component of the state estimate matrix defined in Equation 1.1. These models are very important, as they are the heart of the prediction system. They need to cover most, if not all, of the possible scenarios or the predictions will contain more errors.

For more details on how to set up a KF and a detailed explanation of all required mathematical equations, refer to publications such as References 25, 27, and 36.

1.3 Multiple Models Frameworks

Because the dynamics of automobiles can vary over time, the stated Equations 1.11 through 1.14 are already defined to capture the different states in which a vehicle can be found, but these independent state equations need to be merged to produce only one prediction. There are several algorithms that exist to modify the stochastic information, and they are well known for their ability to automatically adapt the filter in real time to match any variation of the errors involved.

One such algorithm is termed multiple models adaptive estimation (MMAE). The MMAE algorithm runs several Kalman filters in parallel, each operating using different dynamic or stochastic models. The MMAE algorithm is used to select a single "best" Kalman filter solution, or the algorithm can be used to combine the output from all the Kalman filters in a single solution. A possible limitation of such an approach would be the large computational burden imposed by running multiple Kalman filters. However, with improved processor technology, such an approach can now be considered even for real-time applications [25].

Another such algorithm is the IMM estimation, which, even though it works in a similar manner as the MMAE by running multiple Kalman filters in parallel, is more mathematically involved and takes into account the probability of the next KF selection, making it more accurate than the MMAE in many scenarios.

As the dynamic state of vehicles is highly variable over time, the model selected has to meet the conditions of very different situations. However, a solution based on the implementation of a unique model that fulfills the consistency requirements of scenarios with high dynamic changes provokes unrealistic noise considerations when mild maneuvers are performed, diminishing the filter efficiency and impoverishing the final solution. Therefore, two different interactive multimodel filters have been developed and implemented to identify which one is better for this type of scenario.

The MMAE algorithm is used to select a single "best" Kalman filter solution, or the algorithm can be used to combine the output from all the Kalman filters in a single solution. It uses only the previous evaluation of the individual filters used to identify which one should be used in the calculation of the next estimated location.

The IMM algorithm calculates the probability of occurrence for each of the individual filters and uses that information to identify which of the filters will be more predominant. This algorithm continues calculating the probability for each of the steps throughout the whole run; therefore, the IMM should be more accurate than the MMAE.

1.3.1 Multiple Models Adaptive Estimation

The classic MMAE uses a bank of m Kalman filters running simultaneously, each tuned to a different dataset. The principle of the MMAE algorithm is described by Figure 1.4, which shows that the new measurements, z_k, are used in a bank of N Kalman filters. Each filter is configured to use either different stochastic matrices or different mathematical models. The updated state estimates, x_{k+1}, for the N Kalman filters are computed using the extended Kalman filter algorithm. The states from each filter are then combined by computing weight factors, and summing the weighted outputs.

There are many different ways in which the weight factors can be computed. The one chosen for this system was the dynamic multiple model method since not one filter will be the correct one at all times. This algorithm is described next.

The weight factors are computed using the recursive formula in Equation 1.15, for N Kalman filters, where $p_n(k)$ is the probability that the nth model is correct. The probability density function, $f_n(z_k)$, is computed for each filter

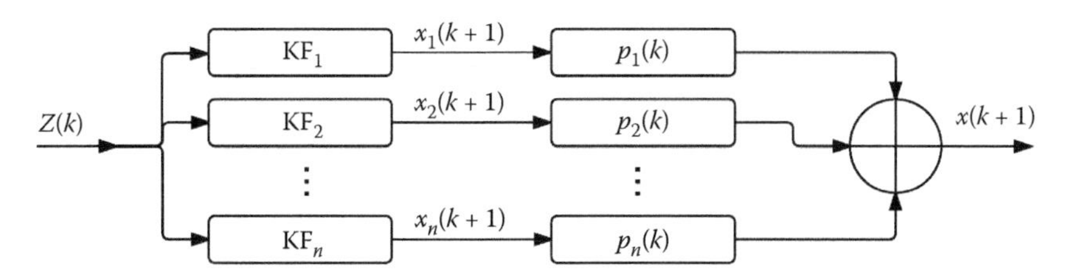

FIGURE 1.4
Multiple model adaptive estimation flow diagram.

based on $V^T \cdot S^{-1} \cdot V$, and its corresponding covariance, S_k, using the formula in Equation 1.17:

$$p_n(k) = \frac{f_n(z_k) \cdot p_n(k-1)}{\sum_{j=1}^{N} f_j(z_k) \cdot p_j(k-1)} \tag{1.15}$$

$$S_k = H \cdot P \cdot H^T \tag{1.16}$$

$$f_n(z_k) = \frac{1}{\sqrt{(2\pi)^{m/2} \mid S_k \mid}} e^{-(1/2)\cdot(V^T \cdot S^{-1} \cdot V)} \tag{1.17}$$

The expression for the covariance in Equation 1.16 reflects the filter's estimate of the measurement residuals, not the actual residuals. This becomes clear when one examines the update expressions for "P" in the Kalman filter: "P" does not depend on the measurement residual. The effect of this is that the expression may indicate some small residual variance, when in fact at particular points in time, the variance is relatively large. This is indeed exactly the case when one is simultaneously considering multiple models for a process—one of the models, or some combination of them, is "right" and actually has small residuals, while others are "wrong" and will suffer from large residuals. Thus, when one is computing the likelihood of a residual for the purpose of comparing model performance, one must consider the likelihood of the actual measurement at each time step, given the expected performance of each model (1.15). This likelihood and probability variables allow the MMEA to determine which one of the filters defined should be used in the estimation of the next location, providing an accurate estimation.

1.3.2 Interacting Multiple Model

The IMM algorithm calculates the probability of occurrence for each of the individual filters and uses that information to identify which of the filters will be predominant. This algorithm continues recalculating the probability for each iteration throughout the whole run, weighting the new probability values against the probability values calculated in the previous iteration, as illustrated in Figure 1.5.

The basic idea of IMM is to compute the state estimate under each possible current model using the four filters defined, with each filter using a different combination of the previous model-conditioned estimates [33]. This method allows coping with the uncertainty on the motion of the automobile by running a set of predefined displacement modes at the same time.

The IMM filter calculates the probability of success of each model at every filter execution, providing combined solution for the vehicle behavior. These

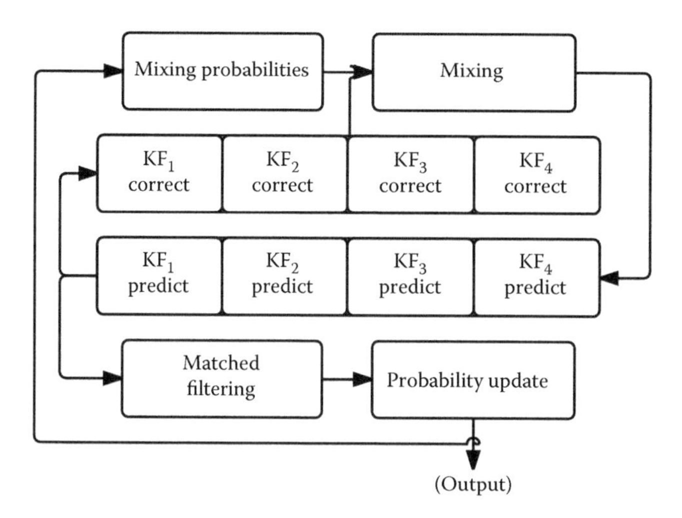

FIGURE 1.5
Interacting multiple models flow diagram.

probabilities are calculated according to a Markov model for the transition between maneuver states, as detailed in Reference 26. To implement the Markov model, it is assumed that at each execution time, there is a probability p^{ij} that the vehicle will make a transition from model state i to state j. Equation 1.18 shows the transition matrix for the four defined KF models defined in Section 1.2.

$$p^{ij} = \begin{bmatrix} CL \rightarrow CL & CL \rightarrow CV & CL \rightarrow CA & CL \rightarrow CJ \\ CV \rightarrow CL & CV \rightarrow CV & CV \rightarrow CA & CV \rightarrow CJ \\ CA \rightarrow CL & CA \rightarrow CV & CA \rightarrow CA & CA \rightarrow CJ \\ CJ \rightarrow CL & CJ \rightarrow CV & CJ \rightarrow CA & CJ \rightarrow CJ \end{bmatrix} \tag{1.18}$$

In Johnson and Krishnamurthy's paper on "An Improvement to the Interactive Multiple Model Algorithm" [34], they describe the IMM as a recursive suboptimal algorithm that consists of five core steps:

- Step 1. Calculation of the mixing probabilities
- Step 2. Mixing
- Step 3. Mode matched filtering
- Step 4. Mode probability update
- Step 5. Estimate and covariance combination

As in any recursive system, the IMM algorithm first needs to be initialized before it can start its four-step recursion. The number of filters used is 4.

Step 1. Calculation of the mixing probabilities

The probability mixing calculation uses the transition matrix (1.18) and the previous iteration model probabilities (1.23) to compute the normalized mixing probabilities (1.19). The mixing probabilities are recomputed each time the filter iterates before the mixing step.

$$\lambda_k(i \mid j) = \frac{p_{ij}\lambda_{k-1}(i)}{\sum_{i=1}^{N} p_{ij}\lambda_{k-1}(i)} \tag{1.19}$$

Step 2. Mixing

The mixing probabilities are used to compute new initial conditions for each of the N filters, four in this case. The initial state vectors are formed as the weighted average of all the filter state vectors from the previous iteration (1.20). The error covariance corresponding to each of the new state vectors is computed as the weighted average of the previous iteration error covariance conditioned with the spread of the means (1.21).

$$x_{k-1}^{oj} = \sum_{i=1}^{N} \lambda_{k-1}(i \mid j)\hat{x}_{k-1}^{i} \tag{1.20}$$

$$P_{k-1}^{oj} = \sum_{i=1}^{N} \lambda_{k-1}(i \mid j) \times \left\{ P_{k-1}^{i} + \left[\hat{x}_{k-1}^{i} - \hat{x}_{k-1}^{0j}\right]\left[\hat{x}_{k-1}^{i} - \hat{x}_{k-1}^{0j}\right]^{T} \right\} \tag{1.21}$$

Step 3. Mode matched filtering

Using the calculated $\hat{x}_{k-1}^{0j}$ and P_{k-1}^{0j}, the bank of four Kalman filters produce outputs $\hat{x}_{k}^{j}$, the covariance matrix P_{k}^{j}, and the probability density function $f_n(z_k)$ for each filter (n) in Equation 1.23, according to the equations for the KF. The covariance for each filter is represented by S_k in Equations 1.22 and 1.23:

$$S_k = H \cdot P \cdot H^{T} \tag{1.22}$$

$$f_n(z_k) = \frac{1}{\sqrt{(2\pi)^{4/2} \mid S_k \mid}} e^{-(1/2)\cdot(V^{T} \cdot S^{-1} \cdot V)} \tag{1.23}$$

Step 4. Mode probability update

Once the new initial conditions are computed, the filtering step (step 3) generates a new state vector, error covariance, and likelihood function for each of the filter models. The probability update step then computes the individual filter probability as the normalized product of the likelihood function and the corresponding mixing probability normalization factor (1.24).

$$\lambda_k(j) = \frac{f_n(z_k)}{\sum_{i=1}^{N} f_n(z_k)} \sum_{i=1}^{N} p_{ij}\lambda_{k-1}(i) \tag{1.24}$$

Step 5. Estimate and covariance combination

This step (1.25 and 1.26) is used for output purposes only; it is not part of the algorithm recursions. Using Equation 1.25, the prediction for $k+1$ can be calculated:

$$\hat{x}_k = \sum_{j=1}^{N} \lambda_k^j \cdot \hat{x}_k^j \tag{1.25}$$

$$P_k = \sum_{i=1}^{N} \lambda_k^j \cdot \left\{ P_k^j + \left[\hat{x}_k^j - \hat{x}_k\right]\left[\hat{x}_k^j - \hat{x}_k\right]^T \right\} \tag{1.26}$$

1.4 Geographic Information System

A GIS is a system for capturing, storing, analyzing, and managing data and associated attributes, which are spatially referenced to the earth. It is a tool that allows users to create interactive queries (user-created searches), analyze the spatial information, edit data and maps, and present the results of all these operations. In this research, we extracted the road information from the maps being used to display the vehicle's location. It is not a very accurate map, but it is enough to demonstrate if the implementation of GIS information with the IMM system improves the prediction of the vehicle's future location or not.

The idea of using GIS data to correct an invalid estimation came about looking at simulations during curves. When the vehicle enters a turn, the prediction of its future locations is very erroneous, many times outside of a road. If the system had a way of knowing the direction of the road ahead, and whether the estimated future location was on an actual road or not, it would be able to correct its estimation and improve its reliability. This is where GIS comes into play, with the assumption that the vehicle will always remain on the road. It is also assumed the driver is handling the vehicle properly and is awake for this GIS correction to be practical. These assumptions, though restrictive, still allow the correction to be useful in scenarios such as road intersections.

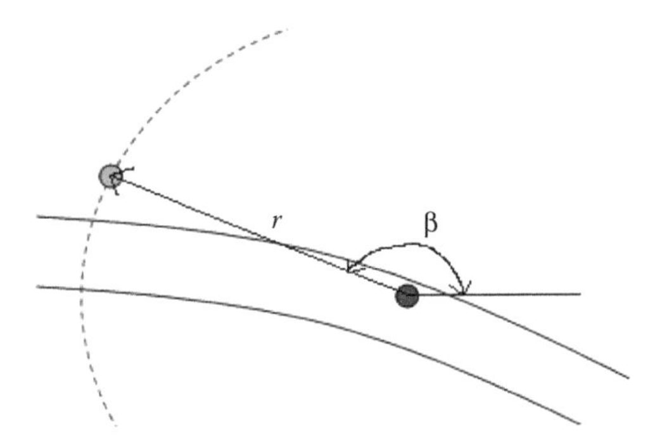

FIGURE 1.6
Displaying parameters used in the method to estimate position on the road.

When a road is designed, the radius of curvature is known, but this information is not available with the GIS data; therefore, a new method is needed to be able to project the estimation outside of the road back in the road.

Because of the limitation of the mapping software used during this research (MapPoint), the only available function to interact with GIS data was to check whether the specific location was on the road or not. A function that provided the distance from the current location to the nearest road would have worked better, but it was not available in MapPoint.

To overcome the limitation described earlier, a method to map the estimated future location outside of the road to an accurate location inside a road had to be designed. From the current GPS location, the distance r and the angle β shown in Figure 1.6 are calculated. The angle β varies with the direction of the movement and is calculated from east being zero degrees. The r is the distance between the current location and the estimated location:

$$\text{Count} = \frac{\text{circumference}}{\text{arc}} \tag{1.27}$$

$$\alpha = \frac{360^{\circ}}{\text{count}} \tag{1.28}$$

The variable arc used in Equation 1.27 is the predefined distance between points in the circumference. The smaller this value, the smaller the increments between checkpoints in the circumference, and the more accurate the measurement. Because the smaller the arc value, the more the points that need to be checked, it required more CPU processing time; so for this research, arc has a value of 0.6 m. This value was selected because the smallest road, even if only a one-way lane, cannot be less than 2 m wide. If we used a value bigger than 2 m, we could have the possibility of missing a road between checkpoints, so we chose a significantly smaller value. The angle α calculated in Equation 1.28 is the actual angle increment needed to match the predefined arc distance on the circumference.

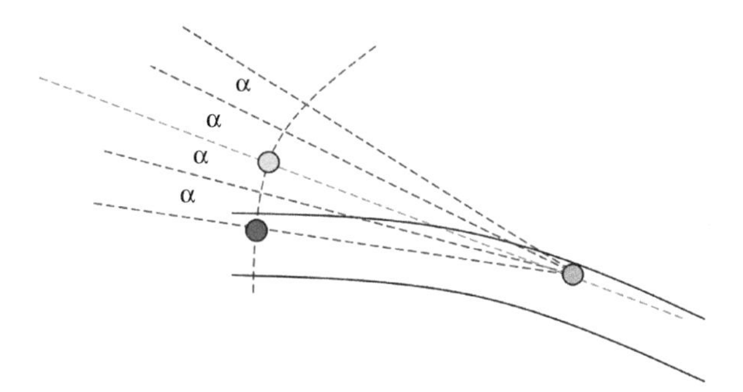

FIGURE 1.7
Geometry used to map estimated future location outside the road to a location inside the road.

With the angle β shown in Figure 1.6 and the angle α calculated in Equation 1.28, running through the checkpoints of the circumference was started. The estimated location is found at angle β and since this estimated location cannot be too far from the actual road, checking was started from this angle β. The system will check both clockwise and counterclockwise increments of α until a point is found on the road. Figure 1.7 provides a graphical view of the GIS error checking implemented. The clockwise and counterclockwise increments will continue to occur until either a road is found and a correction on the estimated future location is made, or a maximum number of increments is reached, and no correction is made. If a correction is made, the new estimated future location will still be the same distance away r; the only difference is its location coordinates.

In Figure 1.8, in MapPoint, the current location is a green dot, the predicted future location is a yellow dot, and the GIS corrected data is a red dot. The smaller red dots are the clockwise and counterclockwise increments described earlier. Visually, in Figure 1.8, the estimated future location is probably incorrect as there is no road in that location. Using GIS data to locate the road, the predicted location can be adjusted to be on the road at the same distance away, as the velocity will probably not change significantly under normal circumstances. The result is a more accurately predicted future location. This method seems to work well during curves, but, as stated earlier, it requires several restrictive assumptions. Therefore, this system could only be useful as a part of a larger and more robust collision avoidance system that took into account some of the scenarios not covered by our proposed method.

1.5 Experimental Results

The experimental setting for testing the models described in Section 1.2 uses a log file of GPS data that contains different scenarios, especially

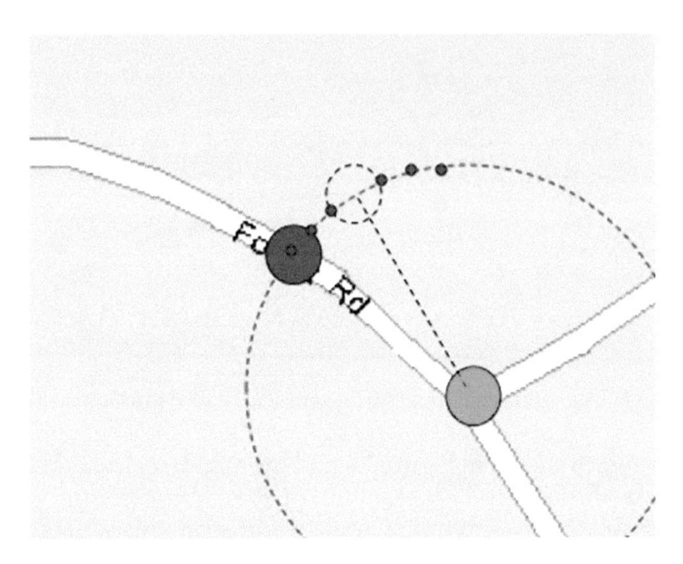

FIGURE 1.8
GIS error correction in MapPoint.

those currently causing problems in existing systems (Figures 1.1 and 1.2). Figure 1.9 shows the trajectory recorded for this research. It has many turns and contains various changes in speed and direction. Because in trying to improve the trajectory estimation during curves, Figure 1.10 shows the curve selected for our experiment.

FIGURE 1.9
Trajectory recorded in GPS log file (Essex Junction, Vermont, USA).

FIGURE 1.10
Selected curve from trajectory recorded.

The selected road curve is definitely a nice sharp turn that occurs at medium speeds (~60 kph). It was felt that this turn would be a good scenario to test the improvements on trajectory estimation.

The code was implemented in Microsoft Visual Basic 6 and Microsoft MapPoint 2004, allowing the software display information in real time on the map as the vehicle moves. Being able to look at the estimated points on an actual map makes it easier to visually inspect and present the system.

Average 3-s ahead estimation errors, shown in the tables in this chapter, are calculated from the difference between estimated and actual GPS locations. It is the distance between the estimated 3-s ahead location and the actual GPS location 3 s later. Actual GPS errors are not accounted for in this research, so both the estimated future location and the actual GPS location should be similarly affected by the GPS error.

1.5.1 Evaluation of SE

To set up a base for the KF evaluation, this experiment defines as simple estimations (SE) the use of the KF models without the KF infrastructure. So, the models defined in Equations 1.11 through 1.14 are run using only the previous data obtained from the GPS.

To evaluate the SE of the individual models and avoid confusion with the KF evaluations, these models are being referenced as simple estimation constant velocity model (SECV), simple estimation constant acceleration model (SECA), and simple estimation constant jerk (SECJ).

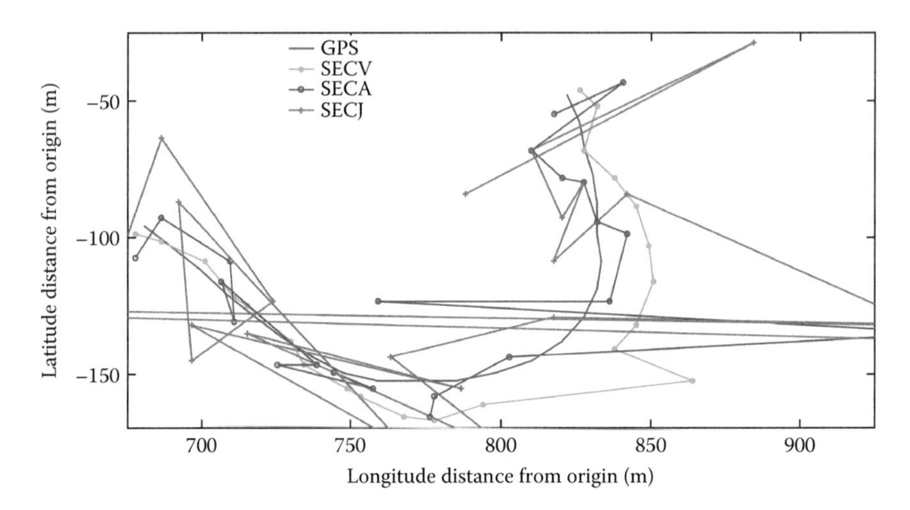

FIGURE 1.11
Comparison of estimated 3-s ahead location and actual GPS readings for all three SE models using the 21 data points for the selected curve.

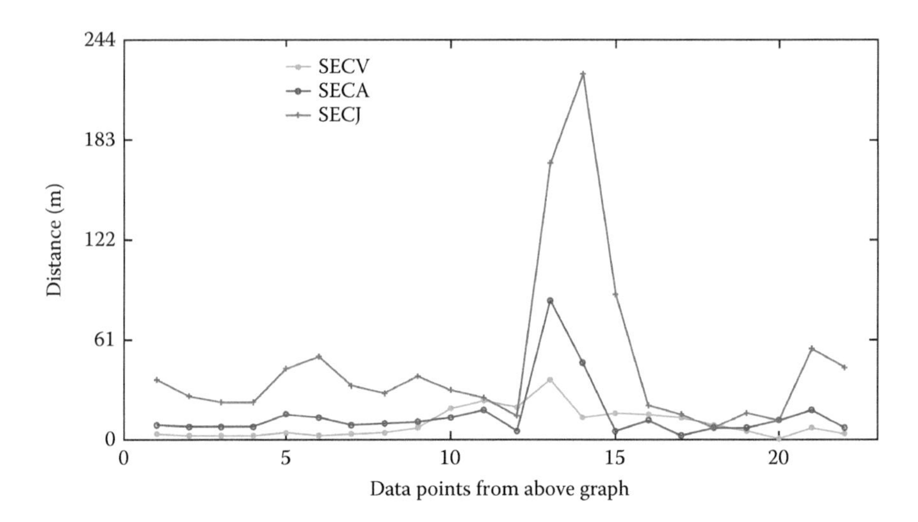

FIGURE 1.12
Calculated error for all SE models between 3-s ahead estimation and actual location 3 s later using the 21 data points for the above specific turn.

Figures 1.11 and 1.12 are two graphical representations of the inaccuracy of the SE models for a 3-s ahead prediction of a vehicle's position. Figure 1.11 is compared to the actual trajectory recorded by the GPS 3 s later, and the most accurate of the three models is the SECV as the curve was taken at an approximately constant velocity. The other two models' estimations have a lot of error because they are assuming the vehicle is moving at constant acceleration and

TABLE 1.1

Average 3-s Ahead Estimation Error for SE Models

	CL (m)	CV (m)	CA (m)	CJ (m)
SE	60.6269	10.6269	16.0947	49.0778

Note: The 21 data points were used for the selected curve in Figure 1.10.

at constant jerk. Figure 1.12 displays the error of each of the models, but showing the number of meters away the 3-s ahead estimated future location is from the actual GPS recorded location 3 s later. Again, the SECV is more accurate than the others in this scenario, but it is still very inaccurate for a reliable collision avoidance system (see Table 1.1 for actual values).

1.5.2 Evaluation of KF

To be able to evaluate, the four KF models, Kalman filter constant location (KF-CL), Kalman filter constant velocity (KF-CV), Kalman filter constant accelerator (KF-CA), and Kalman filter constant jerk (KF-CJ) had to be coded, tested, and tuned individually to get as accurate estimations as are possible. It is given that one of these models will not be very accurate all the time on a real-time GPS log; therefore, in order to calibrate them individually, the GPS log for the full trajectory shown in Figure 1.9 was used to calculate the measurement noise covariance and also each of the process noise covariance for the four models to exercise only one model at the time. To find the values for the process and measurement noise covariance matrices, the data were smoothed out using a moving average window to remove any outlier. The measurement noise covariance was obtained for each of the filters by calculating the covariance of going frame by frame, and calculating the error of the real data to fit into each of the KF models defined in Section 1.2.

Once the filters were tuned, they were individually run through the different scenarios and only the results for the data points in the selected curve were recorded in Table 1.2.

Running the four filters together showed how, when one was very close to the real value, the others were not that accurate. In some instances, more than one filter was accurate, probably when speed changes or acceleration changes were very small. In other cases, none of the four filters was accurate

TABLE 1.2

Average 3-s Ahead Estimation Error for KF Models

	CL (m)	CV (m)	CA (m)	CJ (m)
KF	14.9002	9.8786	7.0812	8.9952

Note: The 21 data points were used for the selected curve in Figure 1.10.

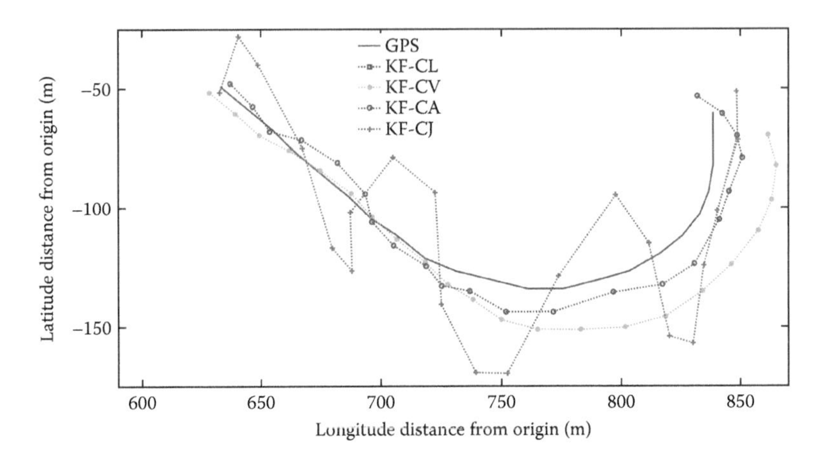

FIGURE 1.13
Comparison of estimated 3-s ahead location and actual GPS reading for all four KF models using the 21 data points for the selected curve.

at all, probably because of an abrupt change in direction or even in speed. The system reads data from the GPS every 1 s, so it is possible, though not common at higher speeds, to have a big change occur during that 1 s, especially in curves. For the most part, 1 s will not allow the speed and direction to change by a big amount (except in some lower-speed scenarios, such as at intersections when making a sharp turn), allowing the filters to estimate the next location somewhat accurately.

From Figures 1.13 and 1.14, we can analyze the results of running the KF individually (each KF is predicting the future location 3 s later in time). Figure 1.13 shows the predicted location 3 s ahead in time on the spatial trajectory (same curve as shown in Figure 1.10), while Figure 1.14 shows the error for each of the predictions of the future vehicle's location 3 s later in time compared to the actual GPS measurement (average estimation error values shown in Table 1.2). Both graphs are needed because KF-CL seems to be accurate in Figure 1.13 because it will always be on a real GPS location since it assumes no movement (constant location). Actually, the KF-CL shows a lot of error in Figure 1.14 since the vehicle was always moving through this curve. The estimated future location for this model will be where the current GPS location is (right over the GPS line), but this will not be accurate if the vehicle is moving, and this is where Figure 1.14 displays this error.

Because the curve selected in Figure 1.10 was driven at a somewhat constant speed, it can be noted that both the KF-CV is the most accurate.

1.5.3 Evaluation of MMAE

The setup of the MMAE framework was considerably more straightforward than the IMM implementation and it seemed to converge all three KF nicely,

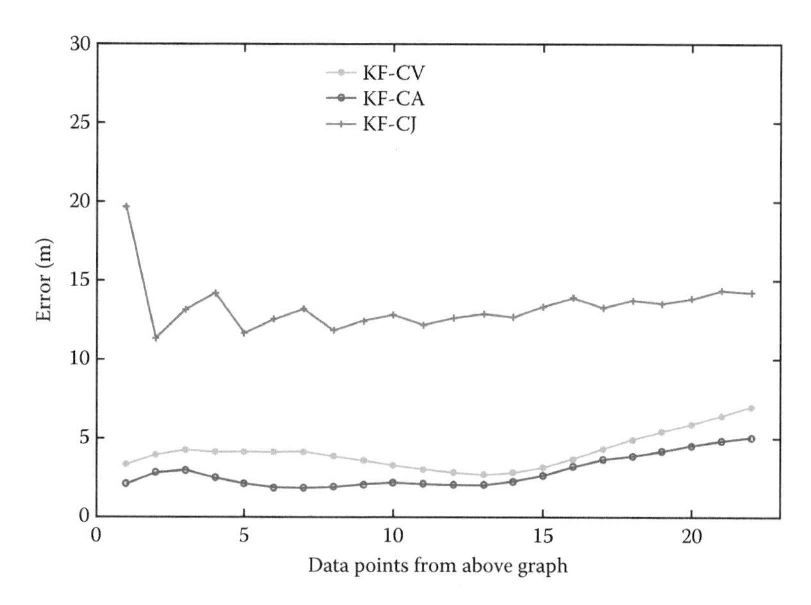

FIGURE 1.14
Calculated error for all KF models between 3-s ahead estimation and actual GPS location 3 s later using the 21 data points for the selected curve.

giving a prediction of a future location closest to the most accurate of the individual KF. Table 1.3 shows the average errors of the 3-s away estimations compared to the actual GPS reading when the vehicle reached that location 3 s later. We can see that this estimated value has too much error to be useful for any type of collision avoidance system; it would just give too many false warnings.

Looking at the MMAE curve in Figure 1.15, we can see that the part of the estimated positions that had the most error was exactly where the turn is, especially at the beginning of it as the vehicle was coming from a straight line and all of a sudden started turning sharply. It takes a few seconds for the system to correct all that error and become more accurate, which makes it not very reliable.

1.5.4 Evaluation of IMM

To set up the IMM, it was necessary to calculate the transition probability matrix in Equation 1.18 using the GPS log for the full trajectory shown in

TABLE 1.3

Average 3-s Ahead Estimation Error for MMAE

Estimated Position	1 s Ahead (m)	3 s Ahead (m)
MMAE	3.4153	10.1924

Note: The 21 data points were used for the selected curve in Figure 1.10.

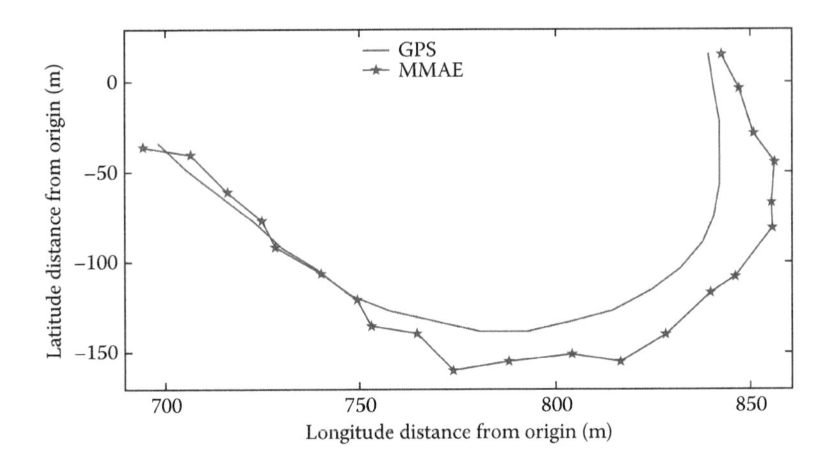

FIGURE 1.15
Comparison of estimated 3-s ahead location and actual GPS readings for the MMAE implementation using the 21 data points for the selected curve.

Figure 1.9. From this full GPS log that contained multiple scenarios, it was determined which transition was occurring frame by frame by comparing the actual measurements from the GPS to the smoothed measurements. The smoothing of the data was done with a rolling window using a combination of median smoothing, splitting the sequence, and Hann's sequence, which removed any abrupt changes from the data. Each transition was determined by the type of change, such as no change, a constant change, and so on. Similarly, by calculating the covariance of the differences in the measurements to each other, the measurement noise covariance matrix (R) was obtained. And last, by calculating the covariance of the differences in the measurements compared to their respective x and y components, the process covariance noise (Q) was obtained for each KF. From this type of information, the below transition probability matrix was obtained:

$$p^{ij} = \begin{bmatrix} 0.154 & 0.154 & 0.385 & 0.308 \\ 0.011 & 0.470 & 0.305 & 0.214 \\ 0.014 & 0.259 & 0.458 & 0.269 \\ 0.002 & 0.243 & 0.508 & 0.247 \end{bmatrix} \tag{1.29}$$

Looking at Equation 1.29, some scenarios are clearly identified. For example, when in a CL state (first row), it is more probable for it to change to a CA or CJ state than to a CV state, and this is understandable because when a vehicle is at a complete stop, to start moving, it will need to accelerate.

Looking at a 1-s ahead estimation allows for some very accurate results but this would not be enough warning time for the driver to react, so this research looks at 3-s away estimations and how accurate they can be. The estimation

TABLE 1.4

Average 3-s Ahead Estimation Error for IMM

Estimated Position	1 s Ahead (m)	3 s Ahead (m)
IMM	2.9056	8.7880

Note: The 21 data points were used for the selected curve in Figure 1.10.

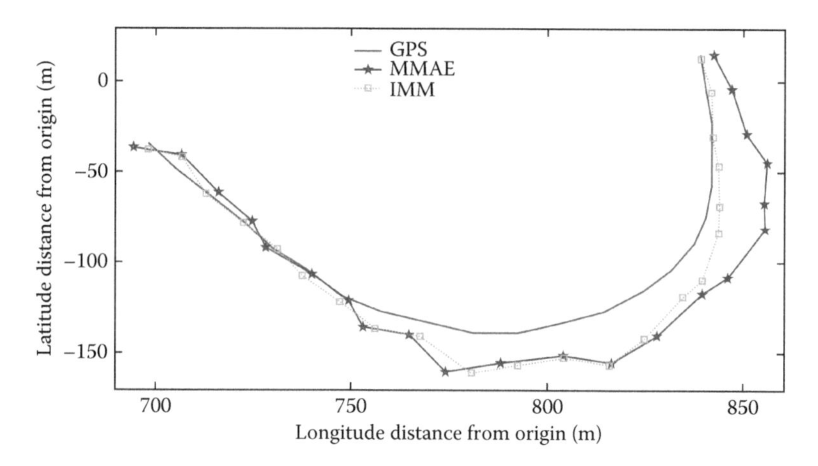

FIGURE 1.16
Comparison of estimated 3-s ahead location and actual GPS readings for the MMAE and IMM implementations using the 21 data points for the selected curve.

average errors measured from running the IMM setup explained earlier is displayed in Table 1.4 and in Figure 1.16, comparing the estimated 3-s ahead positions against the GPS values.

Looking at Figure 1.16, it can be observed that the IMM had a lot of error at the beginning of the turn and, after a few seconds, converged more with the actual path. So, even though the IMM yields smaller average estimation errors than the MMAE setup, if this method were used as part of a collision avoidance system, it would still produce many false warnings.

1.5.5 Evaluation of the GIS Error Correction

In Figure 1.17, the frame shots of the simulation program can be seen. It shows in light yellow the three positions corresponding to 1- and 3-s away estimations. In red, the images show the corrected predicted location for each of the 1- and 3-s away estimations.

The implementation of GIS data with the IMM estimation process showed very promising results. It is easy to see in Figure 1.17 how much the GIS correction helps with the actual estimation of future positions of the vehicle. To look at some numbers, Table 1.5 can be used to confirm this

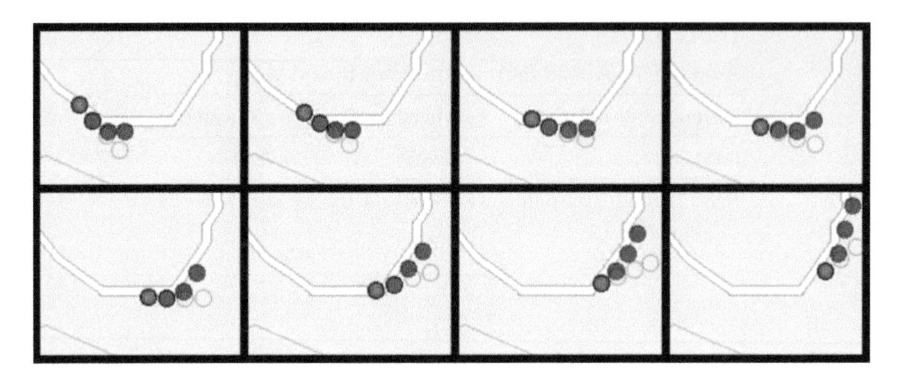

FIGURE 1.17
Frame shots of simulation during the selected curve.

visual conclusion. This table shows the average error for the selected turn and we can see a noticeable difference compared to the method without any GIS correction, especially when looking at the 3 s ahead. A 45% improvement was identified when using the GIS correction method, but the error is still significant when predicting the vehicle's location 3 s ahead of time. Figure 1.18 is another visual aid to be able to compare it to the previous two methods and see how much more accurate this is.

Table 1.6, similarly to Table 1.5, shows the average errors for the estimated future vehicle's location 1 and 3 s ahead in time, but the whole trajectory as shown in Figure 1.9 was used to test this system. The numbers do not show as great an improvement as in Table 1.5 because, when the vehicle is traveling in a straight line, the error in the estimated future location is smaller, and therefore adding GIS correction is not as beneficial.

The GIS error correction method used in this research is somewhat simple and straightforward. It can possibly be improved with other existing methods, but it was enough to help determine whether using GIS data with the trajectory estimation models was an improvement as easily observed in Figure 1.19. Overall, even though GIS does show to be very helpful, especially during curves, it is still not enough to use it by itself, as it was set up for this research. A much-needed improvement would be the implementation of more sensors, which could run at higher frequencies, to contribute to more accurate estimations.

TABLE 1.5
Average 3-s Ahead Estimation Error for IMM+GIS

Estimated Position	1 s Ahead (m)	3 s Ahead (m)
IMM	2.9056	8.7880
IMM+GIS	1.7834	4.8244

Note: The 21 data points were used for the selected curve in Figure 1.10.

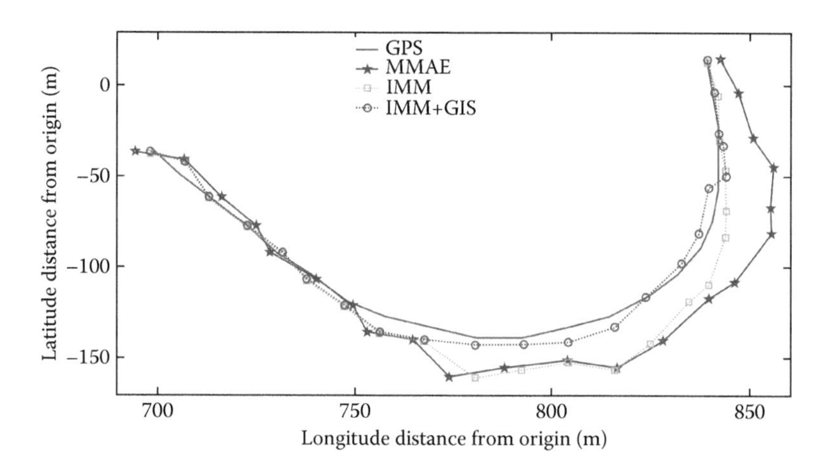

FIGURE 1.18
Comparison of 3-s ahead estimated location between MMAE, IMM, IMM+GIS, and actual GPS locations 3 s later using the 21 data points for the selected curve.

TABLE 1.6
Average 3-s Ahead Estimation Error for IMM+GIS

Estimated Position	1 s Ahead	3 s Ahead
IMM	1.9461	6.5276
IMM+GIS	1.8872	5.1423

Note: Used the 800 data points for whole trajectory in Figure 1.9.

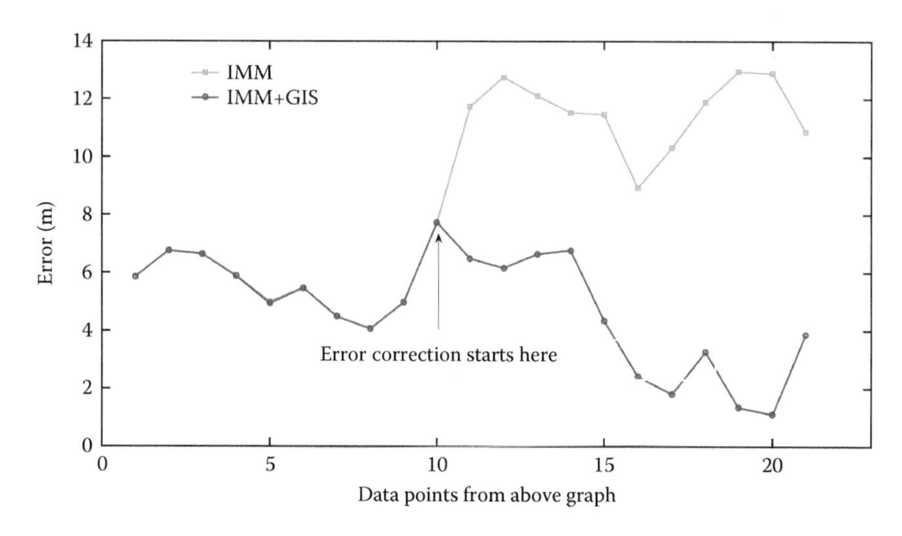

FIGURE 1.19
Error measured between the 3-s ahead IMM estimation (with and without GIS) and the actual GPS readings 3 s later using the 21 data points for the selected curve.

1.6 Conclusions

This chapter implemented four KFs to account for the identified possible states an automobile can be found in (constant location, constant velocity, constant acceleration, and constant jerk). These four KFs were set up to be part of an IMM system that provided the predicted future location of the automobile up to 3 s ahead in time. To improve the prediction error of the IMM setup, this study added an iterated geometrical error detection method based on the GIS system. The assumption made was that the automobile would remain on the road, so predictions of future locations that fall outside of the road were corrected accordingly, making great reduction to prediction error, as shown in the experimental results.

The research observed estimation values at 3 s ahead in time to allow for enough reaction time if this setup were to be used in some type of driver's aid system. As shown in this research, a 3-s ahead estimation has a lot of error, but, with the help of GIS data, this error can be reduced drastically, especially during turns, which is where research seems to have the most problems with [10].

The idea of merging spatial GPS data with GIS road information, given some assumptions, has proven to improve the accuracy of predicting a vehicle's future. And, in some scenarios, it could be an interesting addition to a collision avoidance system.

Despite the improved predictions shown in this chapter, this system can be further improved. The implemented GIS method in this study was straightforward and could be improved by looking into more detailed GIS data and being able to determine the lane the vehicle is driving in to correct with more accuracy a bad estimated future location. The spatial data used from the GPS can also be complemented by using other types of sensors less error prone and that can run at a frequency higher than 1 Hz.

References

1. A. Tascillo, R. Miller, An in-vehicle virtual driving assistant using neural networks, *Proceedings of the International Joint Conference on Neural Networks*, Vol. 3, pp. 2418–2423, Portland, Oregon, July 20–24, 2003.
2. M. Lee, Y. Kim, An efficient multitarget tracking algorithm for car applications, *IEEE Transactions on Industrial Electronics* 50(2), 2003: 397–399, doi: 10.1109/TIE.2003.809413.
3. A. Amditis, E. Bertolazzi, M. Bimpas et al. A holistic approach to the integration of safety applications: The INSAFES subproject within the European Framework Programme 6 Integrating Project PReVENT, *IEEE Transactions on Intelligent Transportation Systems* 11(3), 2009: 554–566, doi: 10.1109/TITS.2009.2036736.

4. Y. Ikemoto, Y. Hasegawa, T. Fukuda, K. Matsuda, Zipping, weaving: Control of vehicle group behavior in non-signalized intersection, *IEEE Proceedings of International Conference on Robotics and Automation*, Vol. 5, pp. 4387–4391, New Orleans, Louisiana, April 26–May 1, 2004.
5. S.G. Wu, S. Decker, P. Chang, T. Camus, J. Eledath, Collision sensing by stereo vision and radar sensor fusion, *IEEE Transactions on Intelligent Transportation Systems* 10(4), 2009: 606–614, doi: 10.1109/TITS.2009.2032769.
6. S. Pietzsch, T.D. Vu, J. Burlet, O. Aycard, T. Hackbarth, N. Appenrodt, J. Dickmann, B. Radig, Results of a precrash application based on laser scanner and short-range radars, *IEEE Transactions on Intelligent Transportation Systems* 10(4), 2009: 584–593, doi: 10.1109/TITS.2009.2032300.
7. M. Chowdhary, Driver assistance applications based on automotive navigation system infrastructure, *Proceedings from ICCE International Conference in Consumer Electronics*, pp. 38–39, Los Angeles, California, June 18–20, 2002.
8. A. Drane, C. Rizos, *Positioning Systems in Intelligent Transportation Systems*, Artech House Inc., Norwood, Massachusetts, 1998.
9. R. Bishop, *Intelligent Vehicle Technology and Trends*, Artech House Inc., Norwood, Massachusetts, 2005.
10. J. Ueki, J. Mori, Y. Nakamura, Y. Horii, H. Okada, Development of vehicular-collision avoidance support system by inter-vehicle communications, *Proceedings of IEEE 59th Vehicular Technology Conference*, Vol. 5, pp. 2940–2945, Milan, Italy, May 17–19, 2004.
11. R. Miller, H. Qingfeng, An adaptive peer-to-peer collision warning system, *Proceedings of IEEE 55th Vehicular Technology Conference*, Vol. 1, pp. 317–321, Birmingham, Alabama, May 6–9, 2002.
12. A.P. Zhang, J. Gu, E. Milios, P. Huynh, Navigation with IMU/GPS/digital compass with unscented Kalman filter, *Proceedings of the IEEE International Conference on Mechatronics & Automation*, pp. 1497–1502, Niagara Falls, Canada, July 29–August 1, 2005.
13. B. Barshan, H.F. Durrant-Whyte, Inertial navigation systems for mobile robots, *IEEE International Transactions on Robotics and Automation* 2(3), 1995: 328–342, doi: 10.1109/70.388775.
14. R. Toledo, M.A. Zamora, B. Ubeda, A.F. Gomez-Skarmeta, An integrity navigation system based on GNSS/INS for remote services implementation in terrestrial vehicles, *IEEE Proceedings from the Intelligent Transportation Systems Conference*, pp. 477–480, Washington DC, October 3–6, 2004.
15. R. Toledo, M.A. Zamora, B. Ubeda, A.F. Gomez, High integrity IMM-EKF based road vehicle navigation with low cost GPS/INS, *IEEE Transactions on Intelligent Transportation Systems ITISFG* 8(3), 2007, doi: 10.1109/TITS.2007.902642.
16. W. Kim, G. Jee, J. Lee, Efficient use of digital road map in various positioning for ITS, *IEEE Transactions on Position Localization and Navigation*, pp. 170–176, San Diego, California, 2000, doi: 10.1109/PLANS.2000.838299.
17. X. Zhang, Q. Wang, D. Wan, Map matching in road crossings of urban canyons based on road traverses and linear heading-change model, *IEEE Transactions on Instrumentation and Measurement* 56(6), 2007: 2795–2803, doi: 10.1109/TIM.2007.908116.
18. Chang Bok Lee, Dong Doo Lee, Nak Sam Chung, Ik Soo Chang, E. Kawai, F. Takahashi, Development of a GPS codeless receiver for ionospheric calibration and time transfer, *IEEE Transactions on Instrumentation and Measurement* 42(2), 1993: 494–497, doi: 10.1109/19.278610.

19. M. Matosevic, Z. Salcic, S. Berber, A comparison of accuracy using a GPS and a low-cost DGPS, *IEEE Transactions on Instrumentation and Measurement* 55(5), 2006: 1677–1683, doi: 10.1109/TIM.2006.880918.
20. M.E. Najjar, P. Bonnifait, A roadmap matching method for precise vehicle localization using belief theory and Kalman filtering, *IEEE 11th International Conference in Advanced Robotics*, Coimbra, Portugal, June 30–July 3, 2003.
21. Y. Cui, S.S. Ge, Autonomous vehicle positioning with GPS in urban canyon environments, *IEEE Transactions on Robotics and Automation* 19(1), 2003: 15–25, doi: 10.1109/TRA.2002.807557.
22. Holux Technology Inc., Holux GR-213 GPS Specifications, http://www.holux.com/JCore/UploadFile/7011686.pdf.
23. A. Lahrech, C. Boucher, J.C. Noyer, Fusion of GPS and odometer measurements for map-based vehicle navigation, *IEEE Proceedings from the International Conference on Industrial Technology*, Vienna, Austria, December 8–10, 2004.
24. S. Sukkarieh, Low cost, high integrity, aided inertial navigation systems for autonomous land vehicles, PhD thesis, University of Sydney, Australia, 2000.
25. C. Hide, T. Moore, M. Smith, Multiple model Kalman filtering for GPS and low-cost INS integration, *Proceedings of ION GNSS*, 2004, Long Beach, CA, September 21–24, 2004.
26. J. Bohg, Real-time structure from motion using Kalman filtering, Technische Universitat Dresden, March 2005.
27. G. Welch, G. Bishop, An introduction to the Kalman filter, SIGGRAPH, 2001, Course notes.
28. C. Hu, W. Chen, Y. Chen, D. Liu, Adaptive Kalman filtering for vehicle navigation, *Journal of Global Positioning Systems* 2(1), 2003: 42–47.
29. Y. Zhang, H. Hu, H. Zhou, Study on adaptive Kalman filtering and algorithms in human movement tracking, *Proceedings of the IEEE International Conference on Information Acquisition*, Hong Kong, China, July 5–10, 2005.
30. L.C. Yang, J.H. Yang, E.M. Feron, Multiple model estimation for improving conflict detection algorithms, *IEEE Proceedings from the Conference on Systems, Man and Cybernetics*, Vol. 1, pp. 242–249, The Hague, Netherlands, October 10–13, 2004.
31. X. Wang, Maneuvering target tracking and classification using multiple model estimation theory, PhD dissertation, University of Melbourne, 2001.
32. A. Derbez, B. Remillard, The IMM CA CV Performance, unpublished.
33. Y. Bar-Shalom, X.R. Li, T. Kirubarajan, *Estimation with Applications to Tracking and Navigation*, Wiley and Sons, Hoboken, New Jersey, 2001.
34. L.A. Johnson, V. Krishnamurthy, An improvement to the interactive multiple model (IMM) algorithm, *IEEE Transactions on Signal Processing* 49(12), 2001: 2893–2908, doi: 10.1049/ip-rsn:19971105.
35. M. Green, How long does it take to stop? Methodological analysis of driver perception-brake times, *Transportation Human Factors* 2(3), 2000: 195–216.
36. H. Himberg, Y. Motai, Head orientation prediction: Delta quaternions versus quaternions, *IEEE Transactions on Systems Man and Cybernetics Part B-Cybernetics* 39(6), 2009: 1382–1392, doi: 10.1109/TSMCB.2009.2016571.

2

*Intelligent Forecasting Using Dead-Reckoning with Dynamic Errors**

2.1 Introduction

Sensor fusion and tracking techniques have potential applications for the vehicle and the infrastructure as introduced in Reference 1, something we can appreciate from the industrial sensing intelligence found in the ITS area [2]. The overall function of ITS is to improve decision making, often in real time, improving the operation of the entire transport system. This can range from systems with intelligent route planning implemented to avoid some specific type of traffic from certain areas [3], to registering the position of vehicle-borne sensors for infrastructure assessment [4], to systems designed to prevent collisions between the users [5]. This research could fall under the latter category.

A collision avoidance system is only as good as its accuracy in warning the driver—either human or automated. An accurate system will minimize the number of false warnings so the driver takes them seriously. Designing the architecture of this type of system involves using many sensors, for intelligent control and decision making, and finding the right balance between the number of sensors implemented, type, and their overall contributions to the intelligent forecasting system.

There are mainly two types of designs for a collision avoidance system. Self-sufficient systems are those that can obtain enough information by themselves, such as those in References 6 through 8. Interactive systems are those that interact with the infrastructure or other vehicles to detect a dangerous scenario, such as researched in References 9 through 11 where their systems send spatial information to nearby vehicles to judge the possibility of a collision in the future. While self-sufficient systems are limited to line-of-sight detection, the interactive systems are not limited by this but are more complex. Estimating the future trajectory of a vehicle requires multiple sensors that need to be merged together and put through a set of prediction models.

* This chapter is a revised version of the author's paper in IEEE Transactions on Industrial Informatics. DOI: 10.1109/TII.2015.2514086, approved by IEEE Intellectual Property Rights.

Multi-Sensor Data Fusion (MSDF) techniques are required to combine and process data [12,13]. This has been traditionally performed by some form of Kalman [14] or Bayesian filters [15]; however, in recent years, there has been a trend toward the use of soft techniques such as fuzzy logic and artificial neural networks [16,17]. Furthermore, there can be two ways of setting up an MSDF system: centralized or decentralized. A centralized approach suffices for most common scenarios where the sensors are synchronous, but a decentralized approach is convenient when the sensors should be treated independently [18–22], as with asynchronous sensors.

In Reference 23, the authors discuss the Optimal Asynchronous Track Fusion Algorithm (OATFA), which evolved from their earlier research on an Asynchronous/Synchronous Track Fusion (ASTF) [24]. They use the IMM algorithm, but replace the conventional KFs with their OATFA. The OATFA treats each sensor separately, passing the output from each to a dedicated KF, departing from the idea that the best way to fuse data is to deliver it all to a central fusion engine. The paper's IMM-OATFA results show position estimation errors that are about half of conventional IMM setups. However, as pointed out by the same authors in Reference 25, all measurement data must be processed before the fusion algorithm is executed. Similarly, the authors of Reference 26 create asynchronous holds, where, from a sequence of inputs sampled at a slow sampling rate, it generates a continuous signal that may be discretized at a high sampling rate. Despite the benefits of making the asynchronous system into a synchronous one by using these methods, restrictions are observed, where, if for some reason a sensor is delayed in providing its data or is offline for a few cycles, the whole system needs to wait.

In References 27 through 29, the authors also look into problems of getting measurements from multiple sensors, but they focus on measurements being out of sequence and not on missing measurements. Therefore, while this is a very important topic on some scenarios, for the system that was used in this study, having all the sensors being processed locally, it will be assumed that all measurements are in the correct sequence, and there should not be a reason for some of them getting out of sequence.

Another method to fuse asynchronous sensors is discussed in Reference 30. In this paper, the authors synchronize the output of the sensors by estimating the data of the slower sensors for the time stamps where no data were available from them. Even though the method used to estimate the unavailable readings is very rudimentary (based only on the previous reading), this idea allows the system to run at the fastest frequency of its sensors. This difference, compared to the previously referenced papers, allows the system to make any corrections as soon as data are available, making its estimations more accurate in some scenarios.

The contribution of this paper is a dead-reckoning (DR) system that runs at the frequency of its fastest sensor to update its prediction as soon as a change is detected. The difference from other DR implementations, subject to cumulative errors, is that our dead-reckoning with dynamic error (DRWDE) continually updates the noise covariance matrices when any sensor remains offline by innovating a dynamic Q matrix in the KFs. This constant modification of the true weight of each measurement helps to counteract the cumulative error of the DR when the measurements are estimated and not real.

2.2 Position Estimation Techniques

The KF [31] was first proposed in the 1960s and has been shown to be a form of Bayesian filter [32]. From a series of noisy measurements, the KF is capable of estimating the state of the system in a two-step process: correct and then predict [33–35], as illustrated in Figure 2.1.

The KF has a long history of accurately predicting future states of a moving object and has been applied to many different fields [36–39], including transportation, which is why it was chosen for this research. The elements of the state vector used (x) are the position, velocity, and acceleration of the vehicle, all available from the different sensors. Keep in mind that the position (x_v) and velocity (v_v) components of the state estimate have an x and y component to them (east-west and north-south directions), and the acceleration (a_v) has an n and t component to it (normal and tangential acceleration). So, the state vector matrix will be $X = (s_x, s_y, v_x, v_y, a_n, a_t)$.

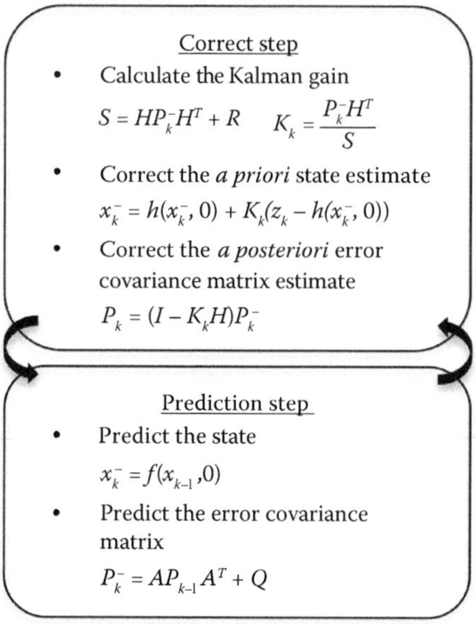

Correct step

- Calculate the Kalman gain

  $S = HP_k^- H^T + R \quad K_k = \frac{P_k^- H^T}{S}$

- Correct the *a priori* state estimate

  $x_k^- = h(x_k^-, 0) + K_k(z_k - h(x_k^-, 0))$

- Correct the *a posteriori* error covariance matrix estimate

  $P_k = (I - K_k H)P_k^-$

Prediction step

- Predict the state

  $x_k^- = f(x_{k-1}, 0)$

- Predict the error covariance matrix

  $P_k^- = AP_{k-1}A^T + Q$

FIGURE 2.1
The Kalman filter two-step recursive process.

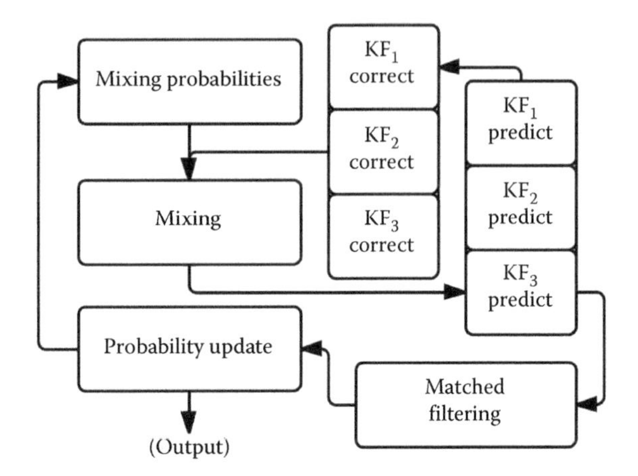

FIGURE 2.2
Flowchart for the three KFs in an IMM framework.

The estimated error covariance (P) for the state estimate is based on the relationships between each of the elements to the others. The error covariance matrix is a dataset that specifies the estimated accuracy in the observation errors between all possible pairs of vertical levels.

Together with P, the Jacobian matrix of the measurement model (H), the measurement noise covariance (R), and the measurement noise (σ_m) are used to calculate the Kalman gain (K). Once the K is calculated, the system looks at the measured data (Z) to identify the error of the predicted position obtained from the dynamic models defined, and uses it to adjust P.

The IMM framework was used in this system [44], as shown in Figure 2.2. It can calculate the probability of success of each model at every filter execution, providing combined solution for the vehicle behavior [40–42]. These probabilities are calculated according to a Markov model for the transition between maneuver states, as detailed in Reference 43. To implement the Markov model, it is assumed that at each execution time, there is a probability p^{ij} that the vehicle will make a transition from model state i to state j.

2.3 DRWDE Using Kalman Filters

This system uses three different sensors: a Garmin 16HVS GPS receiver and Fugawi 3 GPS navigation software, an AutoEnginuity OBDII ScanTool (which obtains the velocity from the vehicle's internal system), and a Crossbow 3-axis accelerometer. This set of sensors offers data at different rates (asynchronous) and also at the same rates (synchronous); one measurement from two of the sensors overlap (homogeneous) but most of them do not (heterogeneous). The accelerometer measures normal and tangential acceleration every tenth of a second, the ScanTool measures velocity every 1 s, and the GPS measures position, velocity, and heading every 1 s (timing precise).

A problem with some of the existing research, as mentioned in Section 2.1, is that sensors can unexpectedly go offline and not provide data when expected. The system in this study will need to handle this without slowing down the running frequency of the overall system and then wait for the sensor to come back online. This in turn means that the system can run at the frequency of its fastest sensor. If the system can continue to run and handle the missing data, it will allow for a quicker correction of the estimation if a change occurs in the spatial movement of the vehicle.

2.3.1 System Architecture

In this setup, the GPS sensor provides the location (s_x,s_y), the velocity (v), and the angle of direction (β) using north as the zero as shown in Figure 2.3. Then the ScanTool sensor provides the velocity (v), and the accelerometer provides normal acceleration (a_n) and tangential acceleration (a_t) as illustrated in Figure 2.4.

The jerk j (acceleration change) in this study's equations is included as the factor responsible for the noise in the measurements; therefore, the jerk term is represented as the prediction noise (σ_p). The different linear dynamic models to be used in the KF used in this research are defined as shown in Equations 2.1 through 2.3.

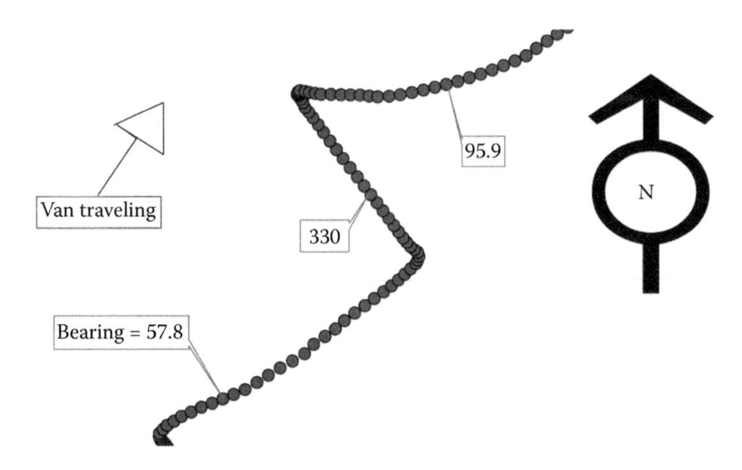

FIGURE 2.3
Bearing measurements.

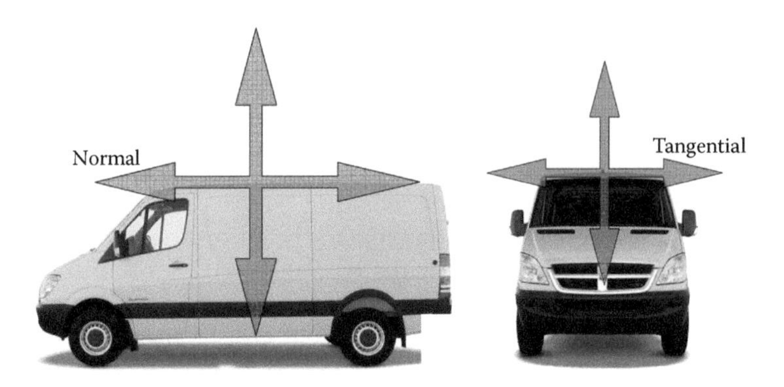

FIGURE 2.4
Accelerometer measurements.

Constant Location (CL)

$$\begin{aligned} s(k) &= s(k-1) + \sigma_{p_s} \\ v(k) &= 0 \\ a(k) &= 0 \end{aligned} \tag{2.1}$$

Constant Velocity (CV)

$$\begin{aligned} s(k) &= s(k-1) + v(k-1) \cdot \Delta k + \sigma_{p_s} \\ v(k) &= v(k-1) + \sigma_{p_v} \\ a(k) &= 0 \end{aligned} \tag{2.2}$$

Constant Acceleration (CA)

$$\begin{aligned} s(k) &= s(k-1) + v(k-1) \cdot \Delta k + \frac{1}{2} a(k-1) \cdot (\Delta k)^2 + \sigma_{p_s} \\ v(k) &= v(k-1) + a(k-1) \cdot \Delta k + \sigma_{p_v} \\ a(k) &= a(k-1) + \sigma_{p_a} \end{aligned} \tag{2.3}$$

In the flow of this setup (Figure 2.5), when a sensor goes offline and the data needed for the models are not present, for example, velocity, the missing data are derived from the data obtained by the remaining online sensors, making this estimation more accurate than only using the offline previous reading of the sensor to estimate what would be its current value. This is insufficient, however, as the longer a sensor remains offline, the more the noise accumulated in the estimation of its value, which in turn affects the overall prediction of the future spatial location of the vehicle. To handle this properly, we have to modify dynamically the noise covariance matrices.

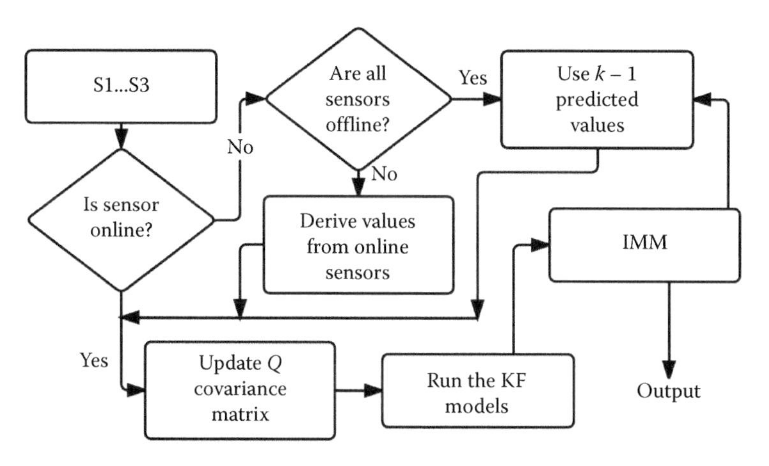

FIGURE 2.5
Flowchart of our DRWDE system.

2.3.2 *Q* Matrix in the KF

The process noise covariance (Q) of the KF is defined based on the estimated prediction noise (σ_p). A simple approach to estimating Q is using an extensive dataset of common scenarios. For this system, because this research wanted to be able to handle sensors going offline at any given time and for any given period of time, an innovative method was devised that makes the Q matrix dynamic, allowing the noise to vary depending on the number of iterations the different variables go through without getting an actual measurement from the corresponding sensor.

2.3.2.1 Mathematical Framework for Improvement

A discrete and dynamic linear system and measurement equation can be generally expressed as shown in Equation 2.4, where k is the current instance and $k+1$ is the future instant for which the data are being estimated. For the linear discrete system, x is the state, F is the state transition function, and B is the control input function. And for the measurement equation, Z is the sensor reading, H is the sensor function, and n is the sensor noise covariance.

$$\begin{aligned} x_{k+1} &= F_k \cdot x_k + B_k \cdot u_k + w_k \\ Z_{k+1} &= H_{k+1} \cdot x_{k+1} + n_{k+1} \end{aligned} \tag{2.4}$$

Given the intermediate data for the instant t_k between the instances when all sensors are online (t_i and t_{i+1}), it is possible to make a prediction for the instant t_{k+1} based on the data at instant t_k, which will most probably result in a better prediction than if using the instant t_i. There are two possible approaches to handle the missing data when sensors are offline.

The first option is to fill in the missing measurements with those of $\hat{x}_{k+1}$, which is the prediction to the intermediate instant. The risk for the minimal quadratic error [45] is $(\hat{x}-E(x))^t \cdot M(\hat{x}-E(x))+trace[MP]$, where M is the defined positive matrix of the quadratic error, and $P = E\langle(\hat{x}-E(x))\cdot(\hat{x}-E(x))^t\rangle$, with the corresponding reduction of the actual measurements when sensors are online.

The second option is to calculate, with the current data obtained from the online sensors, the noise errors for the given small time interval, and obtain a better approximation of the missing measurements, with the goal of obtaining a better Q covariance matrix.

In the first option, the greater the interval $\hat{x}$ and $E(x)$, the error will generally be greater. In the second option, there may not always be a relationship that will yield a good estimation, but experimental runs can help evaluate this approach to determine if the estimation is indeed better.

Because $x_{i+1} = \left[I + F_{k+1}(t_{i+1} - t_{k+1})\right] + w'_{k+1}(t_{i+1} - t_{k+1})$ can also be written as $x_{i+1} = [I + F_{k+1}(t_{i+1} - t_{k+1})]x_{k+1} + w'_{k+1}(t_{i+1} - t_{k+1}) - [I + F_{k+1}(t_{i+1} - t_{k+1})]u_{k+1}$, the corresponding process covariance matrix will be $m = w'_{k+1}(t_{i+1} - t_{k+1}) - \left[I + F_{k+1}(t_{i+1} - t_{k+1})\right]u_{k+1}$. If φ_l is the vector formed by the elements of row l from F_{k+1}, $|\varphi_l \cdot u_{k+1}| \le \|\varphi_l\| \cdot \|u_{k+1}\|$, and if we operate at

$$E\left[m_{k+1}^t m_{k+1}\right] \le E\left[\|w'_i\|^2\right](t_{i+1} - t_{k+1})^2 + E\left[\|u_{k+1}\|^2\right]\left[1 + \left(\sum \|u_{k+1}\|\right)(t_{i+1} - t_{k+1})\right]^2 \quad (2.5)$$

we can define the trace of the covariance matrix of the process as $E[w'^t_i w'_i](t_{i+1} - t_{k+1})^2 = E[\|w'_i\|^2](t_{i+1} - t_{k+1})^2$, which will show an improvement when

$$E\left[\|w'_i\|^2\right](t_{i+1} - t_{k+1})^2 + E\left[\|u_{k+1}\|^2\right]\left[1 + \left(\sum \|u_{k+1}\|\right)(t_{i+1} - t_{k+1})\right]^2 < E\left[\|w'_i\|^2\right](t_{i+1} - t_i)^2 \quad (2.6)$$

which can be rewritten as

$$E\left[\|u_{k+1}\|^2\right] < trace(Q_i) \cdot \frac{1 - \left(\dfrac{t_{i+1} - t_{k+1}}{t_{i+1} - t_i}\right)^2}{\left[1 + \left(\sum \|\varphi_l\|\right)(t_{i+1} - t_{k+1})\right]^2} \quad (2.7)$$

As shown above, a smaller trace of the Q matrix would suppose a general improvement of the covariance matrix of the process, and, therefore, the resulting estimation. However, if a sufficiently general condition is required, then there is a need to study the matrix $E\left[m_{k+1} m_{k+1}^t\right]$ for each specific case, where m_k represents the status of a specific sensor at a given instant.

To approximate the unknown magnitudes, if $x = (x_0 x_1 \ldots x_n)^t$ verifies that $x_{l+1} = \dot{x}_l \ \forall l = 0, 1, \ldots, n-1$, and x_n is the function we have for known measurements in the intermediate instances, it is possible to approximate any x_p for $p = 0, 1, \ldots, n-1$ through a Taylor polynomial representation with its respective error (2.5).

$$x_p(t) = x_p(t_i) + \frac{x_{p+1}(t_i)}{1!}\Delta t_k + \cdots + \frac{x_n(t_i)}{(n-p)!}(\Delta t_k)^{n-p} + \int_{t_i}^{t} \frac{1}{(n-p)!}\dot{x}_n(t_i - y)^{n-p}\,dy \quad (2.8)$$

The measurements of the variables will have an error. Given $\bar{x}_l$ the obtained measurement of x_l, the corresponding error $\varepsilon_l = \bar{x}_l - x_l$, in this first step, is due to $w(t_i)\Delta t_k$. Then, we can accordingly modify the Taylor polynomial as shown below:

$$\begin{aligned} x_p(t_j) = {} & x_p(t_i) + \frac{x_{p+1}(t_i)}{1!}\Delta t_k + \cdots + \frac{x_n(t_i)}{(n-p)!}(\Delta t_k)^{n-p} \\ & - \sum_{m=p}^{n} \frac{\varepsilon_m(t_i)}{(m-p)!} + \int_{t_i}^{t} \frac{1}{(n-p)!}\dot{x}_n(t_i - y)^{n-p}\,dy \end{aligned} \quad (2.9)$$

With this procedure, the measurement $\bar{x}_p(t_i)$ of $x_p(t_i)$ will have an error of

$$\varepsilon_j = \sum_{m=p}^{n} \frac{\varepsilon_m(t_i)}{(m-p)!} + \int_{t_i}^{t} \frac{1}{(n-p)!}\dot{x}_n(t_i - y)^{n-p}\,dy \quad (2.10)$$

If the function of which we have known measurements in $t_j \in (t_i, t_{i+1})$ is x_0, then

$$\begin{aligned} x_{p+1}(c) = \dot{x}_p(c) = {} & \frac{\bar{x}_p(t_k) - \bar{x}_p(t_i)}{\Delta t_k} \\ = {} & \bar{x}_{p+1}(t_i) + \frac{\bar{x}_{p+2}(t_i)}{2!}(\Delta t_k)^1 + \cdots + \frac{\bar{x}_n(t_i)}{(n-p)!}(\Delta t_k)^{n-p-1} \\ & - \frac{1}{\Delta t_k}\left[\varepsilon_p(t_k) - \sum_{m=p}^{n} \frac{\varepsilon_m(t_i)}{(m-p)!} + \int_{t_i}^{t} \frac{1}{(n-p)!}\dot{x}_n(t_i - y)^{n-p}\,dy\right] \end{aligned} \quad (2.11)$$

For a given $c \in (t_i, t_k)$, we can approximate $x_{p+1}(t_k)$ as

$$x_{p+1}(t_j) = \bar{x}_{p+1}(t_i) + \frac{\bar{x}_{p+2}(t_i)}{2!}(\Delta t_k)^1 + \cdots + \frac{\bar{x}_n(t_i)}{(n-p)!}(\Delta t_k)^{n-p-1} \quad (2.12)$$

with an error of

$$\varepsilon_{p+1}(t_j) = x_{p+1}(c) - x_{p+1}(t_k) + \frac{1}{\Delta t_k}\left[\varepsilon_p(t_k) - \sum_{m=p}^{n} \frac{\varepsilon_m(t_i)}{(m-p)!} + \int_{t_i}^{t} \frac{1}{(n-p)!}\dot{x}_n(t_i - y)^{n-p} dy\right] \tag{2.13}$$

where the difference $x_{p+1}(c)-x_{p+1}(t_j)$, which depends on the stability of $x_{p+1}(t)$, is expected to be lower as Δt_k is small.

With the new data obtained from online sensors, the process can be repeated for the next intermediate instances t_{k+1}, t_{k+2}; which, in general, the error will continue to increase as the time gap increases. The exact value of the errors will be unknown in general, so this research will have to be bounded through statistical estimations; even though, in reality, the actual implementation, and not the theoretical validation of the formulas, will determine if there is an improvement in the estimations. For this, it must be taken into account that, due to the cumulative error accumulated with each iteration, an excessive number of iterations will be counterproductive, and will make the estimations worse.

In the case that the function of the more frequently known measurements x_q is not x_0 or x_n, it will suffice to consider on the one hand $(x_0 \ldots x_q)$, and on the other hand $(y_0 \ldots y_m) = (x_0 \ldots x_m)$, and proceed with each group accordingly. If there were more functions with known measurement data, in general, the remaining would be estimated using the closest one, or one of the closest ones.

2.3.2.2 Dynamic Process Noise Covariance (Q)

In the case when all the sensors are available, the formulas for the CA models will depend on the location, velocity, and acceleration measurements in a given instant, and will also depend on the prediction noise σ_p. In this case, σ_p is based on the jerk (j), which will have a variable and unknown value. Based on the *Lagrange form of the remainder* of Taylor's formula, there is a value for j that will yield the exact measurements. Therefore, to set an upper bound of the expected value (E), it suffices to identify an upper bound for j, and calculate the corresponding integrals to obtain each E. But, because in a real-time execution of the system, all the values of j are not known ahead of time, this research made it a moving range so the system can dynamically tune itself.

In summary, to determine $Q = E[\sigma_p \sigma_p^T]$, this research starts by defining j_k (acceleration change) as the least upper bound (supremum) of the dataset collected so far, that is, $\max\left\{|j_{t_k}|, |j_{n_k}|, |j_{t_{k-1}}|, |j_{n_{k-1}}| \cdots |j_{t_{k0}}|, |j_{n_{k0}}|\right\}$.

If the state vector defined in Section 2.2 and the Kalman models defined in Section 2.3.1 are used, and if j_n is to the right of j_t, and for the CA model (2.3), $x(k) = F(k) \cdot x(k-1) + \sigma_p$ has Equation 2.14:

$$x(k)=\begin{bmatrix}1 & \Delta k & \frac{1}{2}(\Delta k)^2\\ 0 & 1 & \Delta k\\ 0 & 0 & 1\end{bmatrix}\cdot x(k-1)+\begin{bmatrix}\frac{1}{6}j(\Delta k)^3\\ \frac{1}{2}j(\Delta k)^2\\ j(\Delta k)\end{bmatrix} \tag{2.14}$$

Furthermore, in this system, it will also take into account the error in the estimations for location, velocity, and acceleration when the sensor providing the corresponding value is offline, and consider for how long it has been offline.

Now, given $M_{k'}(x)$ as the total measurement error of a variable x such that in the step when all sensors are online, $m = 0$, and in the following m step(s), only the accelerometer sensor is online. Because sensors can go offline independently of each other, a different m is needed to identify each sensor: m_1 for the GPS sensor, m_2 for the ScanTool sensor, and m_3 for the accelerometer.

Therefore, this research can now define Q for the CA model as shown in Equation 2.15:

$$Q_{CA}=\begin{bmatrix}M(s)^2 & M(sv) & M(sa)\\ M(vs) & M(v)^2 & M(va)\\ M(as) & M(av) & M(a)^2\end{bmatrix} \tag{2.15}$$

So, each of the process error elements in the Q matrix can be derived. For the position elements (x/y), it is obtained that

$$E\left[M^2(s)\right]\le\frac{(\Delta k)^6}{36}\sum_{i=0}^{m_1-1} j_{k-i}^2 \tag{2.16}$$

The proof of the expected value calculations for each prediction noise σ_p element in the process noise covariance (Q) matrix to show how to arrive at Equation 2.15 starting from Equation 2.14 is shown below:

Derivation for $E\left[M^2(s)\right]$:

$$\begin{aligned}E\left[M_k^2(s)\right] &= E\left[\left(M_{k-1}(s)+\frac{1}{6}j_k(\Delta k)^3\right)^2\right]\\ &= E\left[M_{k-1}^2(s)+2M_{k-1}(s)\frac{1}{6}j_k(\Delta k)^3+\left(\frac{1}{6}j_k(\Delta k)^3\right)^2\right]\\ &= E\left[M_{k-1}^2(s)\right]+2\cdot E\left[M_{k-1}(s)\right]\cdot E\left[\frac{1}{6}j_k(\Delta k)^3\right]+E\left[\frac{1}{36}j_k^2(\Delta k)^6\right]\\ &= E\left[M_{k-1}^2(s)\right]+E\left[\frac{1}{36}j_k^2(\Delta k)^6\right]\end{aligned} \tag{2.17}$$

and

$$
\begin{aligned}
E\left[\sigma_{p_1}\sigma_{p_1}\right] &= E\left[\sigma_{p_1}^2\right] \\
&= E\left[\left(\frac{1}{6}(j_t \sin\beta + j_n \cos\beta)(\Delta k)^3\right)^2\right] \\
&= \frac{(\Delta k)^6}{36}\cdot\left(E\left[j_t^2 \sin^2\beta + j_n^2 \cos^2\beta\right] + 2E[j_t]E[j_n]E[\sin\beta + \cos\beta]\right) \\
&\le \frac{(\Delta k)^6}{36}\cdot\left(E\left[j^2 \sin^2\beta + j^2 \cos^2\beta\right] + 2\cdot 0\cdot 0\cdot E[\sin\beta + \cos\beta]\right) \\
&= \frac{j^2(\Delta k)^6}{36}\cdot E\left[j_t^2 \sin^2\beta + j_n^2 \cos^2\beta\right] = \frac{j^2(\Delta k)^6}{36}
\end{aligned} \tag{2.18}
$$

Therefore,

$$
\begin{aligned}
E[M_k^2(s)] &\le \frac{1}{36} j_k(\Delta k)^2 + \frac{1}{36} j_{k+1}(\Delta k)^2 + E\left[M_k^2(s)\right] \\
&= \frac{1}{36} j_k^2(\Delta k)^2 + \ldots + \frac{1}{36} j_{k+1}^2(\Delta k)^6 + E\left[\sigma_{p_1}^2\right] \le \frac{(\Delta k)^6}{36}\sum_{i=0}^{m-1} j_{k-i}^2
\end{aligned} \tag{2.19}
$$

Using a similar approach, it was found that for the velocity elements (x/y),

$$
E[M^2(v)] \le \frac{(\Delta k)^4}{4}\sum_{i=0}^{m_2-1} j_{k-i}^2 \tag{2.20}
$$

and, finally, for the acceleration elements (n/t), it was derived that

$$
E[M^2(a)] \le (\Delta k)^2 \sum_{i=0}^{m_3-1} j_{k-i}^2 \tag{2.21}
$$

Also, for the nonzero elements outside of the diagonal, it was calculated that

$$
E[M(s\cdot v)] = E[M(v\cdot s)] \le \frac{(\Delta k)^5}{12}\sum_{i=0}^{m_{1|2}-1} j_{k-i}^2 \tag{2.22}
$$

For a given tangential or normal acceleration, the locations and velocities in the axis directions can be any; therefore, the location and velocity variables are independent of the value of the tangential or normal accelerations. And, similarly, the tangential and normal accelerations are independent of each other. Therefore, the expected value of those errors are zero, and the final Q

matrix that will dynamically increase the corresponding measurement error in relation to how long some sensors (m_i) have been offline (Δk) is defined in Equation 2.23:

$$Q_{CA} = \begin{bmatrix} \frac{(\Delta k)^6}{36} \sum_{i=0}^{m_1-1} j_{k-i}^2 & \frac{(\Delta k)^5}{12} \sum_{i=0}^{m_{1|2}-1} j_{k-i}^2 & 0 \\ \frac{(\Delta k)^5}{12} \sum_{i=0}^{m_{1|2}-1} j_{k-i}^2 & \frac{(\Delta k)^4}{4} \sum_{i=0}^{m_2-1} j_{k-i} & 0 \\ 0 & 0 & (\Delta k)^2 \sum_{i=0}^{m_3-1} j_{k-i}^2 \end{bmatrix} \tag{2.23}$$

Using a similar approach as shown above, this research can derive the dynamic Q matrix for the CV model as shown in Equation 2.24 and for the CL model in Equation 2.25:

$$Q_{CV} = \begin{bmatrix} \frac{(\Delta k)^4}{4} \sum_{i=0}^{m_1-1} a_{k-i}^2 & \frac{(\Delta k)^3}{2} \sum_{i=0}^{m_{1|2}-1} a_{k-i}^2 \\ \frac{(\Delta k)^3}{2} \sum_{i=0}^{m_{1|2}-1} a_{k-i}^2 & (\Delta k)^2 \sum_{i=0}^{m_2-1} a_{k-i}^2 \end{bmatrix} \tag{2.24}$$

$$Q_{CL} = \begin{bmatrix} (\Delta k)^2 \sum_{i=0}^{m_1-1} v_{k-i}^2 \end{bmatrix} \tag{2.25}$$

These Q matrices will be used in the KF prediction step to estimate the error covariance for each of the models. And, as shown in the Q matrices above, the moment a sensor comes back online ($m_i = 0$), the corresponding element in the dynamic Q matrix can be reset to its minimum value.

2.4 Evaluation Criteria

To verify the improvements of using the DRWDE, we will implement and compare the results of the following setups:

- Synchronous sensors using a common KF+IMM implementation (GPS @1 Hz, ScanTool @1 Hz, and Accelerometer @1 Hz)

- Asynchronous sensors using our dynamic DRWDE implementation (GPS @1 Hz, ScanTool @1 Hz, and Accelerometer @10 Hz)

The first setup is to get the IMM working at 1 Hz, which will only run when all sensors are online, therefore, not really using the dynamic part of the Q matrix.

The DRWDE setup is to increase the frequency of the system to 10 Hz to try to take advantage of all the readings from the accelerometer, and try to correct the predictions sooner, instead of having to wait for all the sensors to come back online, as in this first setup. This second setup uses the dynamic Q matrix technique described in Section 2.3.2 to account for the error in the estimation of the data when some sensors are offline.

Using the above two setups helps to track improvements to the overall trajectory prediction when the frequency of the system increases along with the proper handling of the accumulated error in the predictions. If this DRWDE is flexible enough to handle all the different synchronous and asynchronous, homogeneous and heterogeneous data from the sensors in use, improvements should be seen on the predicted future locations, and the system should be able to detect and correct a spatial change in the vehicle much sooner than when the system is forced to run at the speed of its slowest sensor.

The evaluation criteria will be based on comparing the actual prediction errors for both the DRWDE and IMM 1 Hz systems against the true location data obtained from the GPS receiver. Both systems will be run through the same trajectory, and the results looked at in several different ways. First, this research will look at the average prediction error for whole trajectory, but then also separate the trajectory into straight lines, smooth curves, and sharp curves, to better evaluate both systems in the different scenarios. This research will also select one specific smooth curve and one specific sharp curve, and it will look at those results in greater detail, using error histograms and calculating root mean square (RMS) and mean absolute percentage error (MAPE) values using the actual and predicted position S of both systems using the formulas in Equations 2.26 and 2.27, respectively:

$$\mathrm{RMS} = \sqrt{\frac{\sum_{k}^{k'} (S_k - S_{k-3})^2}{(k' - k)}} \tag{2.26}$$

$$\mathrm{MAPE} = \frac{1}{n} \sum_{t}^{n} \left| \frac{A_t - F_t}{A_t} \right| \cdot 100 \tag{2.27}$$

2.5 Experimental Performance of the DRWDE System

2.5.1 Dataset Characteristics

The dataset consists of measurements from the three sensors while driving a vehicle for over 1 hour. The trajectory followed is shown in Figure 2.6, where the vehicle followed the route marked in red.

For this experiment, the GPS sensor takes measurements of the current geographical coordinates in degrees, heading in degrees, and velocity in miles per hour every 1 s. These measurements were converted to meters, radians, and meters per second, respectively.

The ScanTool reads the measurements of the velocity determined by sensors coupled to the wheels of the vehicle in miles per hour every 1 s. This measurement is more accurate than the one obtained from the GPS, so it is used instead of the one from the GPS (except when it is not available).

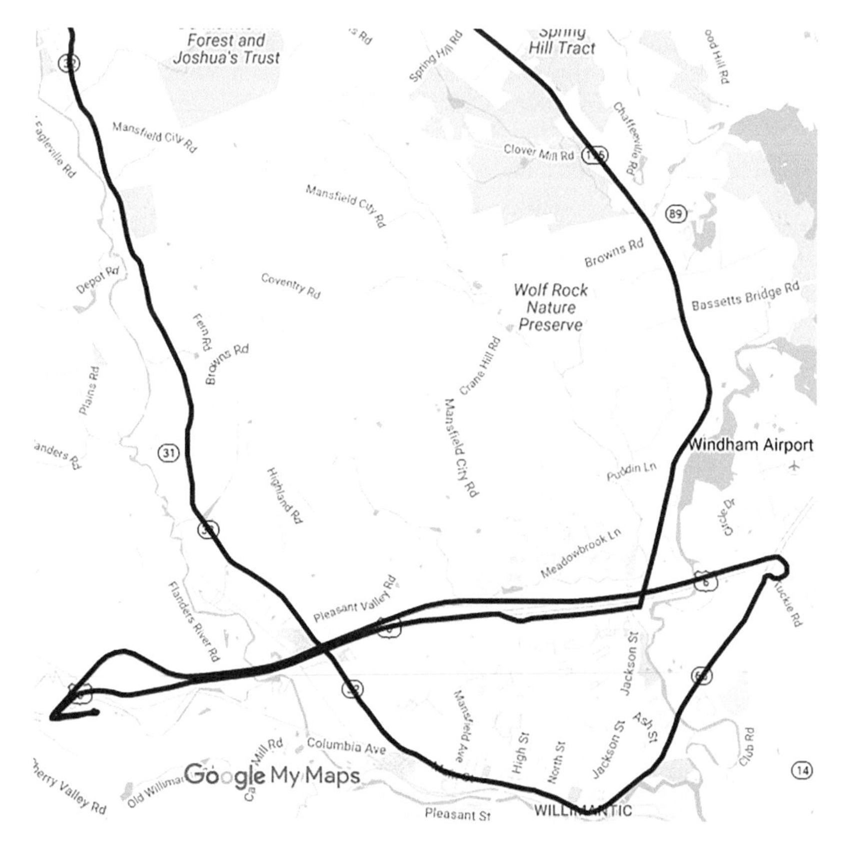

FIGURE 2.6
Map of overall trajectory in Mansfield City, Connecticut (Google maps).

The last sensor used in this experiment is an accelerometer, which takes measurements of the normal and tangential accelerations in volts every 0.1 s. Using a calibration formula provided by the manufacturer of the sensor, the conversion is units to meters per second squared.

The trajectory selected for this research is shown in Figure 2.6. The selected route was selected to include straight and non-straight paths, and also road types driven at different speeds, such as highways, local routes, small streets, and even going through town with several traffic lights and stop signs. This dataset includes most possible scenarios a vehicle could be traveling through.

To be able to create a useful dataset of the data recorded from the trajectory shown in Figure 2.6, this research had to create scripts to map the values from the log file of each sensor to each other, using the time stamp as the common reference between them. In the end, a dataset was created with all the desired measurements in columns, with all available readings in a row for each time stamp. Because only the accelerometer works at 10 Hz, many of the rows only contain acceleration measurements, and this is where the system comes into action and takes advantage of these extra measurements.

For this experiment, the focus was on predicting a trajectory when the vehicle is going through curves, which are the more problematic areas. To be able to evaluate this better, the dataset of the whole trajectory was classified into straight lines, smooth curves, and sharp curves. To determine whether a set of consecutive points in the trajectory was a curve or a straight line, the change in the heading after a period of 2 s was observed; if it was more than 2°, then it was defined as a curve. And, to determine if the curve was a sharp one, the change had to be greater than 10°, otherwise it was defined as a smooth curve. The details of the dataset characteristics for each of these three groups are shown in Tables 2.1 through 2.3.

TABLE 2.1

Central Tendency and Statistical Dispersion Measurements (for Whole Trajectory)

	Distance (m)	Velocity (m/s)	Normal Acceleration (m/s^2)	Tangential Acceleration (m/s^2)
Media	16.71	14.93	–0.51	0.68
Median	16.34	15.20	–0.44	0.69
Minimum value	0.16	0.00	–5.88	–2.26
Maximum value	81.64	34.87	7.06	2.84
Mid-range	40.90	17.43	0.59	0.29
Abs. deviation	4.68	4.48	0.78	0.42
Standard deviation	6.97	5.86	1.10	0.58
Coefficient of variation	0.42	0.39	–2.17	0.84

Note: Whole trajectory dataset was used.

TABLE 2.2

Central Tendency and Statistical Dispersion Measurements (for Smooth Curves)

	Distance (m)	Velocity (m/s)	Normal Acceleration (m/s^2)	Tangential Acceleration (m/s^2)
Media	18.68	17.05	–1.71	0.82
Median	19.04	17.66	–1.69	0.04
Minimum value	0.53	0.89	–5.88	–0.59
Maximum value	28.17	24.72	3.09	2.65
Mid-range	14.35	12.81	–1.40	1.03
Abs. deviation	3.76	3.41	1.50	0.42
Standard deviation	5.58	4.97	1.72	0.53
Coefficient of variation	0.30	0.29	–1.01	0.65

Note: Only dataset classified as ample curves was used.

TABLE 2.3

Central Tendency and Statistical Dispersion Measurements (for Sharp Curves)

	Distance (m)	Velocity (m/s)	Normal Acceleration (m/s^2)	Tangential Acceleration (m/s^2)
Media	10.59	9.36	0.37	1.19
Median	11.07	9.39	–0.15	1.08
Minimum value	0.25	0.00	–2.35	0.20
Maximum value	19.68	17.88	2.94	2.55
Mid-range	9.97	8.94	0.29	1.37
Abs. deviation	5.04	4.33	1.16	0.51
Standard deviation	6.84	5.78	1.34	0.58
Coefficient of variation	0.65	0.62	3.60	0.49

Note: Only dataset classified as sharp curves was used.

TABLE 2.4

Summary of Representative Datasets

	Distance (m)	Velocity (m/s)	Normal Acceleration (m/s^2)	Tangential Acceleration (m/s^2)
Whole trajectory	16.34 ± 6.97	15.20 ± 5.86	–0.44 ± 1.10	0.69 ± 0.58
Smooth curves	19.04 ± 5.58	17.66 ± 4.97	–1.69 ± 1.72	0.04 ± 0.53
Sharp curves	10.59 ± 6.84	9.36 ± 5.78	0.37 ± 1.34	1.19 ± 0.58

Note: Values represent median ± standard deviation of data points used.

Looking at Table 2.4, which only displays the median and standard deviation of the previous three tables, it can be seen that the dataset used for this experiment agrees with how a vehicle would be driven under normal conditions. For example, the standard deviations are not very different from each other for the distance and velocity measured by the sensors, which is expected, as the values do not change much from one point to the next for an average vehicle driving on normal roads. The average for distance and velocity are smaller for the smooth curves than for the sharp curves, which means that the vehicle's speed is more constant through the smooth curves than the sharp curves. The change in movement for sharp curves agrees with how a vehicle would behave in such a scenario, as it will usually have to slow down considerably while turning and then accelerate again as the driver gets a handle on the curve.

Since the main problem with trajectory estimation is during curves based on research reviewed in Section 2.1, this research selected a specific curved scenario from Figure 2.6 and used that dataset to evaluate the DRWDE system and its performance.

The section of the trajectory shown in Figure 2.7 was selected because it has a sharp curve and then a smooth constant curve, which should be a good scenario to test if the system can correct its prediction when the vehicle enters the curve, and maintain it through the whole curve. Sharper curves

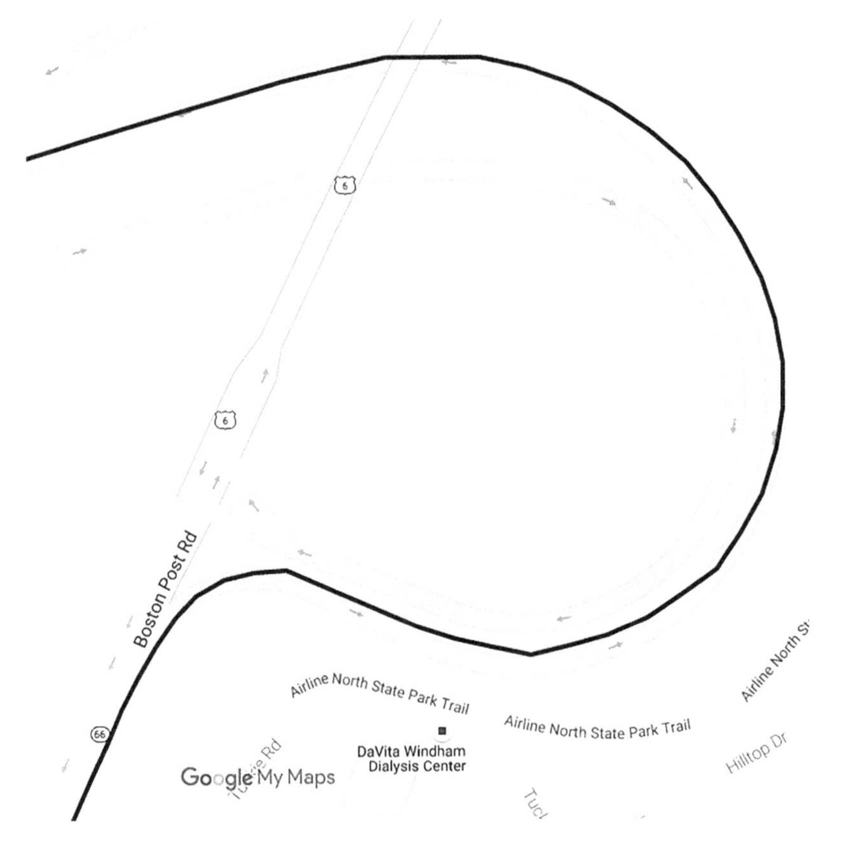

FIGURE 2.7
Map of selected turn for testing (Google maps).

allow our dynamic system to be tested properly as the curve ends up being very short and does not allow a slower system to estimate a trajectory during the actual turn if it only lasts a few seconds. The "selected smooth curve" refers to the longer curve in Figure 2.7 (~30 s of data), and the "selected sharp curve" represents the small curve (bottom left) shown in Figure 2.7 as well (~10 s of data).

The DRWDE setup for this experiment, as explained in Section 2.3, runs at the frequency of its fastest sensor (10 Hz), and uses the dynamic matrices accounting for the accumulated noise of the missing measurements. Also, as mentioned in Section 2.4, data will be running through a common IMM implementation (synchronous sensors) to be able to compare results to the DRWDE setup.

Since the common IMM can only run at the frequency of its slowest sensor, this research defined Δk to be 1 s (1 Hz), and, because all sensors are available during each iteration of the system, this setup does not utilize the dynamic portion of the Q matrix defined in Section 2.3.2.

Also, to properly compare this run to the 10 Hz run, it cannot be assumed the vehicle would move in a straight line between each second, so 10 intermediate points between each second based on the dynamics of the vehicle were defined. This allows us to more accurately compare both runs visually.

Also, the system to estimate a future position of a vehicle will be set up to determine where the vehicle will be 3 s later, which is based on the average human reaction time of 1.5 s to stop a vehicle [46]. In reality, this value would also need to take into account the vehicle's weight and speed to properly determine minimum stopping time necessary.

2.5.2 Evaluation of the Prediction Error

Following the evaluation criteria defined in Section 2.4, the data recorded from the trajectory shown in Figure 2.6 was executed through both systems. The results for the overall trajectory, all smooth and sharp curves, and the selected smooth and sharp curves were recorded in Table 2.5. Keep in mind that the DRWDE is running at 10 Hz, where only the accelerometer

TABLE 2.5

Average Prediction Error

	DRWDE (3 s)	DRWDE (5 s)	IMM 1 Hz (3 s)	IMM 1 Hz (5 s)
Whole trajectory	2.719 ± 2.030	5.844 ± 3.237	3.044 ± 1.800	5.854 ± 3.193
All smooth curves	2.811 ± 1.925	5.633 ± 3.825	2.972 ± 1.737	5.357 ± 3.282
All sharp curves	3.236 ± 2.844	5.063 ± 3.175	4.456 ± 3.307	5.270 ± 3.003
Selected smooth curve	3.051 ± 1.173	5.892 ± 3.200	3.212 ± 1.205	5.483 ± 3.124
Selected sharp curve	2.277 ± 2.388	4.210 ± 1.442	4.090 ± 2.241	5.093 ± 2.981

Note: Values represent median prediction error in meters ± standard deviation of all data points used for both 3- and 5-s ahead predictions.

can provide a measurement at every system iteration, while the GPS and ScanTool provide only 1 reading every 10 iterations, leaving it to our dynamic Q implementation to account for the accumulated error in predicting these missing measurements.

Table 2.5 shows the average prediction errors for both the DRWDE and the IMM 1 Hz run for broader scenarios as well as for our selected curves. If the results of 3-s ahead prediction for the whole trajectory were observed, only a negligible improvement was seen, as expected, since the number of sharp curves in the whole trajectory is very small. Similarly, there is almost no improvement if all smooth curves in the trajectory were observed when compared to the IMM 1 Hz. But since the DRWDE was created to react quickly to changes, it was observed that when taken into account all sharp curves, improvements to the 3-s ahead estimation were seen that are considerably greater for the DRWDE system (3.2 m vs. 4.5 m).

If the focus is now on the selected smooth and sharp curves for the 3-s ahead prediction, the result of the IMM run at 1 Hz is shown in Figure 2.8. The red dotted line shows the predicted location every second (red dots) and the intermediate points derived in between each second (dotted line) to simply show visually what may be happening in between each second.

Also, Table 2.5 shows results for 5-s ahead predictions. As expected, the earlier in time a position is predicted, the more errors there will be, as well as the less reliable the prediction, as shown by the larger standard deviation values for the estimation errors.

Now, for the DRWDE run, Δk was defined to be 0.1 s, which is the period of its fastest sensor (10 Hz). Since only the accelerometer runs at 10 Hz, there

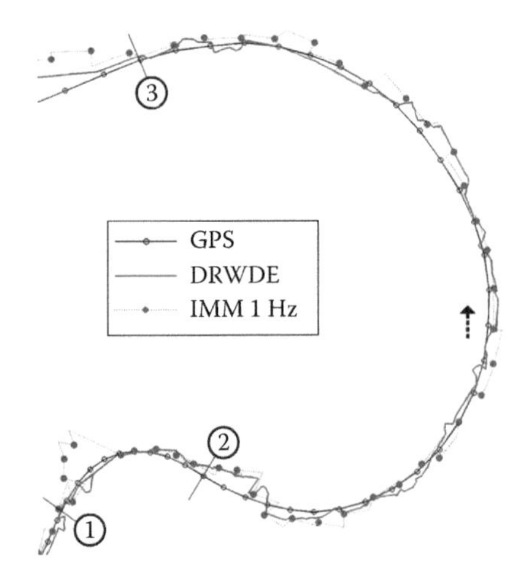

FIGURE 2.8
Comparison between actual path (GPS) and predicted paths by both systems (DRWDE and IMM 1 Hz) for the selected curves. Sharp curve between (1) and (2) and smooth curve between (2) and (3). Direction of movement shown by arrow.

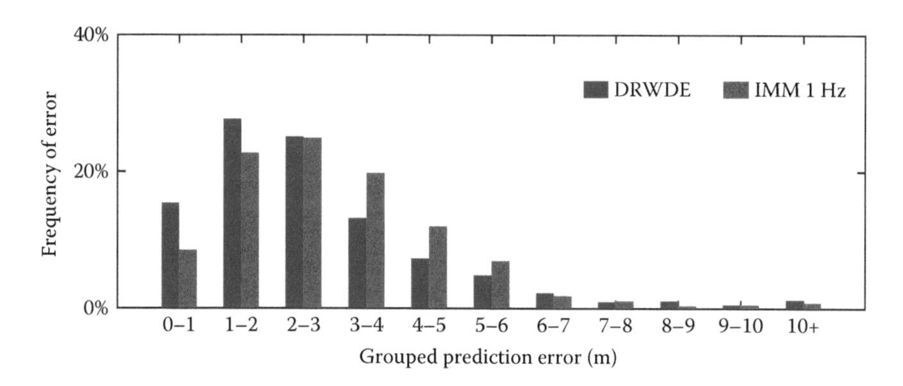

FIGURE 2.9
Frequency of each system's 3-s ahead prediction error for the whole trajectory.

will be many system iterations where the other sensors will be offline, and this is where the dynamic Q variance introduced in Section 2.3.2 comes into play. The result of the DRWDE run at 10 Hz is also shown in Figure 2.8, as the blue solid line.

Figure 2.8 displays the actual trajectory of the vehicle represented by the GPS line, and then the predicted locations 3 s earlier in time by both the IMM 1 Hz run and the DRWDE 10 Hz run (prediction performance is shown later in Figures 2.9 through 2.11). It can be observed that both the 1 Hz and the 10 Hz runs behave somewhat similarly during the smooth curve; this is also represented quantitatively in Table 2.5.

The average error in the predicted locations during the selected smooth curve is only slightly better for the DRWDE (3.0 m vs. 3.2 m). The benefits are clearly seen in the selected sharp curve, where the average error is much lower for the DRWDE (2.3 m vs. 4.1 m). Looking at Figure 2.8, it can be seen that, as the vehicle enters the sharp curve (bottom left), the slower

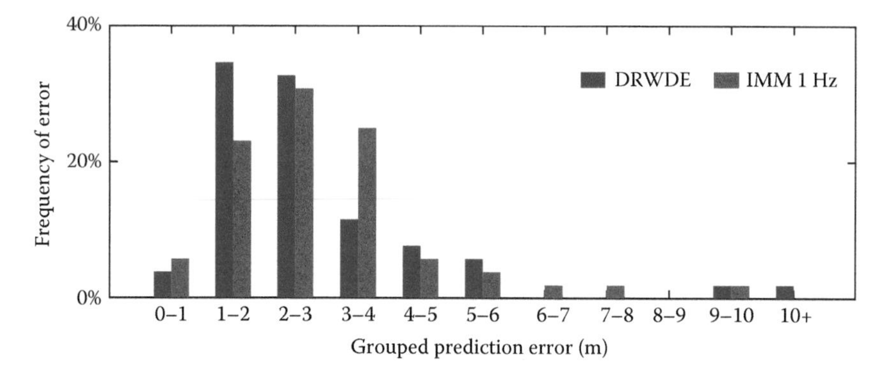

FIGURE 2.10
Frequency of each system's 3-s ahead prediction error for only the smooth curves.

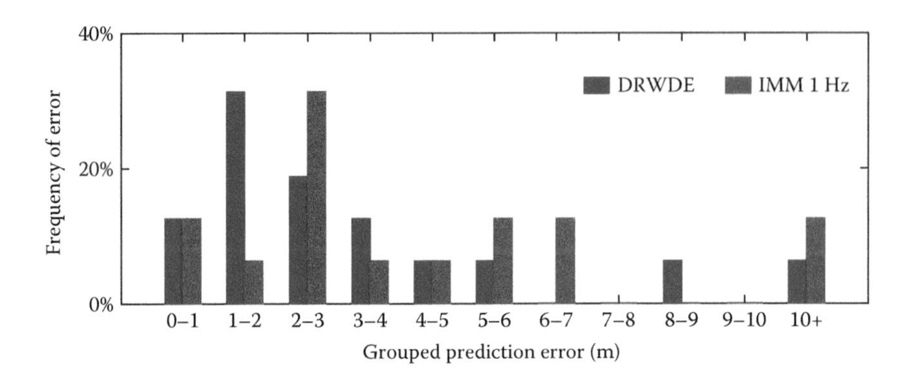

FIGURE 2.11
Frequency of each system's 3-s ahead prediction error for only the sharp curves.

system (red dotted line) is estimating its location to be in more of a straight line, as the vehicle is traveling in a straight line before taking the exit ramp (see Figure 2.7). It can even be seen that there are three red dots (each dot represents 1 s) before the system realizes that the vehicle is turning and can adjust its 3-s ahead prediction accordingly. Looking at the blue line representing the DRWDE run, it can be seen that its line is a lot closer to where the vehicle actually moves through 3 s later in time. The DRWDE 10 Hz system is able to react and correct its future prediction much quicker, using its dynamic covariance matrices to take into account how long a measurement has not been corrected by an actual sensor. As shown in Table 2.5, in the selected sharp curve, a difference of over 1.5 m in accuracy between the two systems can be seen, which is a significant improvement.

2.5.3 RMS and MAPE Error Distribution

A simple visualization of the error distribution for the "whole trajectory," "all smooth curves," and "all sharp curves" prediction errors is shown in Figures 2.9 through 2.11. The charts have the individual prediction errors categorized into groups, where group "0–1" in the x axis contains all the prediction errors that fall between 0 and 1 m, and the y axis shows how frequently the errors fall into each of the groups.

Looking at the histograms in Figures 2.9 through 2.11, it can be observed how the DRWDE system tends to be more often in the first groups, which represent less prediction error. The taller the bars on a given group, the more often the error falls into that error group; therefore, the taller the blue bars on the smaller groups, the more accurate the system.

In Figure 2.9, it can only be seen that the DRWDE outperforms the IMM 1 Hz by a small amount when looking at the overall trajectory, and a larger difference when looking at the results for all smooth curves in Figure 2.10. However, when all sharp curves in Figure 2.11 is observed, a more distinct difference in the prediction accuracy between the DRWDE and the IMM 1 Hz can be seen. To analyze the results for selected smooth and sharp curves specifically, as shown in Figure 2.8, Figure 2.12 was created.

Figure 2.12a represents the error between the estimated future distance the vehicle will travel in the following 3 s, and the actual distance traveled as recorded by the GPS sensor for the selected smooth and sharp curves. Time zero in the figure is set a few seconds before the vehicle enters the sharp curve shown in Figure 2.7. Right at the beginning of the sharp curve, the error in the estimation is quite large for both systems, and that is because the vehicle is moving in a somewhat straight path, so the estimated future position assumes the vehicle will continue to move in the same direction. As soon as the vehicle enters the sharp curve, the first system to detect this change in direction is the DRWDE 10 Hz, as expected, as it can detect this change using the accelerometer, while the GPS is still offline. Once the GPS sensor is back online, the 1 Hz system can also detect this change and can correct its prediction. The upward trend of the lines in Figure 2.12 simply indicates that the vehicle is slowly increasing its velocity and is covering

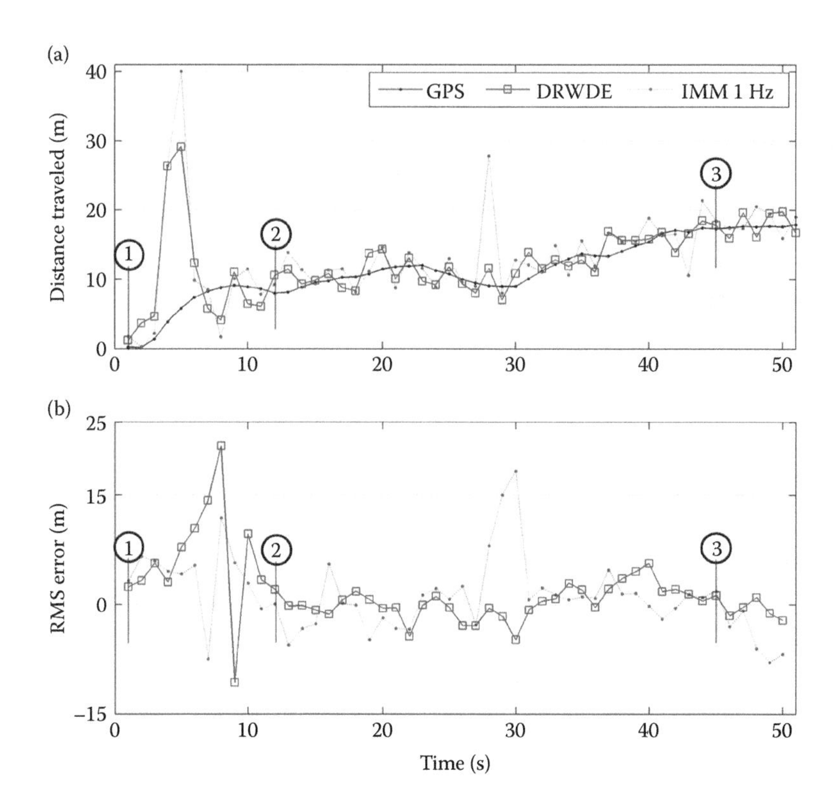

FIGURE 2.12
Position error for a 3-s ahead prediction during the selected curves as shown in Figure 2.8. (a) Actual versus predicted distance traveled per second. (b) RMS error in each prediction.

TABLE 2.6

Mean Absolute Percentage Error

	DRWDE	IMM 1 Hz
Whole trajectory	0.0589	0.0610
All smooth curves	0.0424	0.0417
All sharp curves	0.0987	0.0642
Selected smooth curve	0.0320	0.0333
Selected sharp curve	0.0325	0.0641

Note: Values represent 3-s ahead prediction errors.

more distance in the same period of time (3 s). Only dots at each full second are shown to be able to compare between the two systems.

For a different view of the kind of errors the DRWDE 10 Hz system has, Figure 2.12b was created, which shows the RMS error between the estimated future location (3 s later) and the actual location measured by the GPS. And Table 2.6 shows the MAPE prediction accuracy of this system for the different segment types.

Looking at Figure 2.11 and Table 2.5, it can be concluded that the DRWDE setup really stands out when abrupt changes occur in the movement of the vehicle, and, only then, the fast reaction time shows substantial improvements in the prediction.

2.5.4 Computational Complexity

For completeness, it was also looked into how much of an extra load it is to run the DRWDE system with the dynamic noise matrices compared to the simpler approach of the 1 Hz IMM system. Because the dataset had already been recorded, only the processing time of the system itself was measured. If the processing time of the sensors, especially the accelerometer, is taken into account, the CPU times would be even larger.

Table 2.7 shows different MATLAB® commands used to measure CPU times for each of the systems. All two commands (tic/toc and cputime) measure actual CPU time used by the MATLAB code, but this research is showing both to get a better idea of the accuracy of the measurements. The column tic/toc represents actual start/stop time of execution, while cputime displays

TABLE 2.7

Computational Complexity

	tic/toc (s)	cputime (s)	Data Points	Average Load
DRWDE	389.72	382.88	17,525	1.31
IMM 1 Hz	48.53	47.29	2,187	1.30

Note: Measurements taken on system running through the whole trajectory.

the actual CPU time that was spent executing the code. The system was run on a machine with a dual core 2.0 GHz CPU.

As expected, Table 2.7 shows that the DRWDE 10 Hz system requires a lot more processing power than the simpler IMM 1 Hz system. This is as expected, since the DRWDE system has to handle close to 10 times more data points, and, therefore, yields much longer CPU times. On the same token, looking at the last column, it can be observed that the average load times for every record processed is almost the same for both systems, which shows that the extra computational requirements of the DRWDE's dynamic error processing and measurement noise matrices are not significant at all.

2.6 Conclusions

The key contribution of this research's DRWDE system is the introduction of dynamic noise covariance matrix merged together by an IMM. The longer a sensor remains offline, the less accurate the overall prediction, so the dynamic Q presented in Section 2.3.2 tells the system how true is the value being used.

This DRWDE setup only had three sensors, of which only one of them was running at 10 Hz. The accelerometer is very sensitive to changes in the road, including road bumps; so, relying on this sensor to estimate the values of the other sensors when they were offline had its challenges. However, looking at Section 2.5, it can be concluded that by properly handling the accumulating error for missing measurements, running the system at a higher frequency can yield better predictions, especially when abrupt changes occur. The key here was to be able to accurately account for the accumulating error when sensors go offline and remain offline for an unknown amount of time.

An improvement to this system could be to add more sensors running at high frequencies, for redundancy and to minimize the times sensors are offline. Also, this system could be combined with our previous research [47–48], where the predicted location is compared against GIS to reduce false predictions.

References

1. J.C. Miles, A.J. Walker, The potential application of artificial intelligence in transport, *IEEE Transactions on Intelligent Transport Systems* 153(3), 2006: 183–198, doi: 10.1049/ip-its:20060014.

2. F. Wang, A. Broggi, C.C. White, The road to transactions on intelligent transportation systems: A decade's success, *IEEE Transactions on Intelligent Transportation Systems Magazine* 1(4), 2009: 29–32, doi: 10.1109/MITS.2010.935746.
3. V. Di Lecce, A. Amato, Route planning and user interface for an advanced intelligent transport system, *IET Intelligent Transport Systems* 5(3), 2011: 149–158, doi: 10.1049/iet-its.2009.0100.
4. X. Xu, T. Xia, A. Venkatachalam, D. Huston, The development of a high speed ultrawideband ground penetrating radar for rebar detection, *Journal of Engineering Mechanics* 139(3), 2013: 272–285, doi: 10.1061/(ASCE) EM.1943-7889.0000458.
5. T. Taleb, A. Benslimane, K. Ben Letaief, Toward an effective risk-conscious and collaborative vehicular collision avoidance system, *IEEE Transactions on Vehicular Technology* 59(3), 2010: 1474–1486, doi: 10.1109/TVT.2010.2040639.
6. F. Jimenez, J. Eugenio Naranjo, Improving the obstacle detection and identification algorithms of a laserscanner-based collision avoidance system, *Transportation Research Part C-Emerging Technologies* 19(4), 2011: 658–672, doi: 10.1016/j.trc.2010.11.001.
7. R. Toledo-Moreo, M.A. Zamora-Izquierdo, Collision avoidance support in roads with lateral and longitudinal maneuver prediction by fusing GPS/IMU and digital maps, *Transportation Research Part C-Emerging Technologies* 18(4), 2010: 611–625, doi: 10.1016/j.trc.2010.01.001.
8. L. Jiun-Ren, T. Talty, O.K. Tonguz, A blind zone alert system based on intra-vehicular wireless sensor networks, *IEEE Transactions on Industrial Informatics* 11(2), 2015: 476–484, doi: 10.1109/TII.2015.2404800.
9. R. Toledo-Moreo, M. Pinzolas-Prado, J. Manuel Cano-Izquierdo, Maneuver prediction for road vehicles based on a neuro-fuzzy architecture with a low-cost navigation unit, *IEEE Transactions on Intelligent Transportation Systems* 11(2), 2010: 498–504, doi: 10.1109/TITS.2009.2039011.
10. J. Ueki, J. Mori, Y. Nakamura, Y. Horii, H. Okada, Development of vehicular-collision avoidance support system by inter-vehicle communications, *IEEE 59th Vehicular Technology Conference*, Vol. 5, pp. 2940–2945, Milan, Italy, May 17–19, 2004.
11. H.D. Weerasinghe, R. Tackett, H. Fu, Verifying position and velocity for vehicular ad-hoc networks, *Security and Communication Networks* 4(7), 2011: 785–791.
12. R.C. Luo, C.C. Chang, Multisensor fusion and integration: A review on approaches and its applications in mechatronics, *IEEE Transactions on Industrial Informatics* 8(1), 2012: 49–60, doi: 10.1109/TII.2011.2173942.
13. D. Bruckner, H. Zeilinger, D. Dietrich, Cognitive automation survey of novel artificial general intelligence methods for the automation of human technical environments, *IEEE Transactions on Industrial Informatics* 8(2), 2012: 206–215, doi: 10.1109/TII.2011.2176741.
14. J.B. Gao, C.J. Harris, Some remarks on Kalman filters for the multi-sensor fusion, *Elsevier Information Fusion* 3, 2002: 191–201, doi: 10.1016/S1566-2535(02)00070-2.
15. M. Kafai, B. Bhanu, Dynamic Bayesian networks for vehicle classification in video, *IEEE Transactions on Industrial Informatics* 8(1), 2012: 100–109, doi: 10.1109/TII.2011.2173203.
16. R.R. Murphy, *Sensor Fusion, Handbook of Brain Theory and Neural Networks*, Bradford Book, Cambridge, Massachusetts, 2003.

17. K.Y. Chan, S. Khadem, T.S. Dillon, V. Palade, J. Singh, E. Chang, Selection of significant on-road sensor data for short-term traffic flow forecasting using the Taguchi method, *IEEE Transactions on Industrial Informatics* 8(2), 2012: 255–266, doi: 10.1109/TII.2011.2179052.
18. S.C. Felter, An overview of decentralized Kalman filter techniques, *IEEE Proceedings of the 1990 Southern Tier Technical Conference*, pp. 79–87, Binghamton, NY, April 25, 1990.
19. M. Hua, T. Bailey, P. Thompson et al., Decentralized solutions to the cooperative multi-platform navigation problem, *IEEE Transactions on Aerospace and Electronic Systems* 47(2), 2011: 1433–1449, doi: 10.1109/TAES.2011.5751268.
20. D. Herrero-Perez, H. Martinez-Barbera, Modeling distributed transportation systems composed of flexible automated guided vehicles in flexible manufacturing systems, *Transactions on Industrial Informatics* 6(2), 2010: 166–180.
21. M. Vallee, M. Merdan, W. Lepuschitz, G. Koppensteiner, Decentralized reconfiguration of a flexible transportation system, *IEEE Transactions on Industrial Informatics* 7(3), 2011: 505–516, doi: 10.1109/TII.2011.2158839.
22. H.M. Wang, Q. Yin, X. Xia, Fast Kalman equalization for time-frequency asynchronous cooperative relay networks with distributed space-time codes, *IEEE Transactions on Vehicular Technology* 59(9), 2010: 4651–4658, doi: 10.1109/TVT.2010.2076352.
23. G.A. Watson, T.R. Rice, A.T. Alouani, An IMM architecture for track fusion, *Signal Proceedings of SPIE on Sensor Fusion, and Target Recognition*, pp. 2–13, Orlando, Florida, 2000.
24. A.T. Alouani, T.R. Rice, On optimal asynchronous track fusion, *IEEE Proceedings of 1st Australian Symposium on Data Fusion*, pp. 147–152, Adelaide, South Australia, November 21–22, 1996.
25. G.A. Watson, T.R. Rice, A.T. Alouani, Optimal track fusion with feedback for multiple asynchronous measurements, *Proceedings of SPIE Acquisition on Tracking and Pointing XIV*, Orlando, Florida, July 7, 2000.
26. L. Armesto, G. Ippoliti, S. Longhi et al., Probabilistic self-localization and mapping—An asynchronous multirate approach, *IEEE Transactions on Robotics & Automation Magazine* 15(2), 2008: 77–88, doi: 10.1109/M-RA.2007.907355.
27. Y. Bar-Shalom, H.M. Chen, IMM estimator with out-of-sequence measurements, *IEEE Transactions on Aerospace and Electronic Systems* 41(1), 2005: 90–98, doi: 10.1109/TAES.2005.1413749.
28. X. Shen, Y. Zhu, E. Song et al., Optimal centralized update with multiple local out-of-sequence measurements, *IEEE Transactions on Signal Processing* 57(4), 2009: 1551–1562, doi: 10.1109/ICAL.2008.4636360.
29. X. Shen, E. Song, Y. Zhu et al., Globally optimal distributed Kalman fusion with local out-of-sequence-measurement updates, *IEEE Transactions on Automatic Control* 54(8), 2009: 1928–1934, doi: 10.1109/TAC.2009.2023777.
30. W. Jiangxin, S.Y. Chao, A.M. Agogiono, Validation and fusion of longitudinal positioning sensors in AVCS, *Proceedings of the 1999 American Control Conference*, Vol. 3, pp. 2178–2182, San Diego, California, June 2–4, 1999.
31. R. Kalman, A new approach to linear filtering and prediction problems, *Transaction on ASME* 82, 1960: 34–45, doi: 10.1115/1.3662552.
32. Y. Ho, R. Lee, A Bayesian approach to problems in stochastic estimation and control, *IEEE Transactions on Automatic Control* 9(4), 1964: 333–339, doi: 10.1109/9780470544198.ch58.

33. G. Welch, G. Bishop, *An Introduction to the Kalman Filter*, SIGGRAPH Course Notes, 2001.
34. Y. Bar-Shalom, X.R. Li, T. Kirubarajan, *Estimation with Applications to Tracking and Navigation*, Wiley and Sons, Hoboken, New Jersey, 2001.
35. Y. Bar-Shalom, X.R. Li, *Estimation and Tracking: Principles, Techniques and Software*, Artech House, Norwood, Massachusetts, 1993.
36. B. Mokaberi, A.A.G. Requicha, Drift compensation for automatic nanomanipulation with scanning probe microscopes, *IEEE Transactions on Automation Science and Engineering* 3(3), 2006: 199–207, doi: 10.1109/TASE.2006.875534.
37. P.N. Pathirana, A.V. Savkin, S. Jha, Location estimation and trajectory prediction for cellular networks with mobile base stations, *IEEE Transactions on Vehicular Technology* 53(6), 2004: 1903–1913, doi: 10.1109/TVT.2004.836967.
38. X. Xu, Z. Xiong, X. Sheng, J. Wu, X. Zhu, A new time synchronization method for reducing quantization error accumulation over real-time networks: Theory and experiments, *IEEE Transactions on Industrial Informatics* 9(3), 2013: 1659–1669, doi: 10.1109/TII.2013.2238547.
39. M.H. Kim, S. Lee, K.C. Lee, Kalman predictive redundancy system for fault tolerance of safety-critical systems, *IEEE Transactions on Industrial Informatics* 6(1), 2010: 46–53, doi: 10.1109/TII.2009.2020566.
40. L. Hong, Multirate interacting multiple model filtering for target tracking using multirate models, *IEEE Transactions on Automatic Control* 44(7), 1999: 1326–1340, doi: 10.1109/9.774106.
41. E. Mazor, Interacting multiple model methods in target tracking: A survey, *IEEE Transactions on Aerospace and Electronic Systems* 34(1), 1998: 103–123, doi: 10.1109/7.640267.
42. S.J. Lee, Y. Motai, H. Choi, Tracking human motion with multichannel interacting multiple model, *IEEE Transactions on Industrial Informatics* 9(3), 2013: 1751–1763, doi: 10.1109/TII.2013.2257804.
43. H.A.P. Blom, Y. Bar-Shalom, The interacting multiple model algorithm for systems with Markovian switching coefficients, *IEEE Transactions on Automatic Control* 33(8), 1988: 780–783, doi: 10.1109/9.1299.
44. L.A. Johnson, V. Krishnamurthy, An improvement to the interactive multiple model (IMM) algorithm, *IEEE Transaction on Signal Processing* 49(12), 2001, doi: 10.1049/ip-rsn:19971105.
45. M.S. Grewal, A.P. Andrews, *Kalman Filtering Theory and Practice Using MATLAB*, 2nd edition, p. 132, Wiley-IEEE Press, Hoboken, New Jersey, 2001.
46. M. Green, How long does it take to stop? Methodological analysis of driver perception-brake times, *Transportation Human Factors* 2(3), 2000: 195–216, doi: 10.1207/STHF0203_1.
47. C. Barrios, Y. Motai, Improving estimation of vehicle's trajectory using the latest global positioning system with Kalman filtering, *IEEE Transactions on Instrumentation and Measurement* 60(12), 2011: 3747–3755, doi: 10.1109/TIM.2011.2147670.
48. C. Barrios, Y. Motai, D. Huston, Trajectory estimations using smartphones, *IEEE Transactions on Industrial Electronics* 62(12), 2015: 7901–7910, doi: 10.1109/TIE.2015.2478415.

3

*Trajectory Estimations Using Smartphones**

3.1 Introduction

The overall function of ITS is to improve decision-making, often in real time, improving the operation of the entire transport system. This can go from systems with intelligent route planning implemented to avoid some specific type of traffic in certain areas [1], to keeping track of the position of the vehicle for infrastructure assessment [2], to systems designed to aid with the prevention of collisions between the vehicles [3,4]. For this study, the research focuses on evaluating the use of smartphones as an intermediate step to accelerate the implementation of vehicle-to-vehicle (V2V) and vehicle-to-infrastructure (V2I), which could be used to prevent collisions.

There are two main types of collision avoidance systems: self-sufficient and interactive systems. Self-sufficient systems are those that can obtain enough information from their own sensors, such as those in References 5 through 7, where they placed sensors around the vehicle to maintain a safe following distance or to detect vehicles in the surroundings. Interactive systems are those that, as the name implies, interact with the infrastructure and/or other vehicles, such as researched in References 8 through 10, where their systems send spatial information to nearby vehicles to estimate the probability of a future collision. While self-sufficient systems are limited to line-of-sight detection, the interactive systems account for scenarios farther ahead or even around corners or intersections by predicting and communicating the future estimated trajectories.

The V2V and V2I areas are being well researched these days [11–17], as the government is carefully evaluating the implementation of new technologies to make our roads safer (Figure 3.1). In an article published February 3, 2014 by the United States Department of Transportation (USDOT) [18], the National Highway Safety Administration announced its decision to begin taking the next steps toward implementing V2V technology in all new cars and trucks, after years of research and unprecedented coordination between industry and across government.

* This chapter is a revised version of the author's paper in IEEE Transactions on Industrial Electronics. DOI: 10.1109/TIE.2015.2478415, approved by IEEE Intellectual Property Rights.

FIGURE 3.1
Illustration for V2V and V2I from Reference 18.

When the steps toward implementing V2V technology are defined, car and truck manufacturers will be mandated to enable this in their new vehicles. The challenge faced is that, because V2V relies on other vehicles nearby also supporting V2V technology, there will be a gap of many years when the V2V/V2I will not be able to show its true potential in improving road safety. In an article published by *Forbes* on March 14, 2013 [19], they calculated that the age of the average vehicle on the road is at a record high of 10.8 years, which means there are vehicles on the roads that are 20 years old. Keeping this in mind, it is a long time to wait to ensure full V2V/V2I reliability.

The scientific contribution of this research includes the evaluation of using the smartphone to predict future trajectories for a possible implementation as a temporary hook into the V2V/V2I infrastructure in older vehicles. Allowing drivers of older vehicles the possibility of taking advantage of this new technology would not only benefit them, but it would also benefit the rest of the V2V/V2I-enabled vehicles, as the number of vehicles participating in the system would be much greater. Smartphones are already being used in the transportation field, and one example is the mobile application DriveWell, created by Cambridge Mobile Telematics [20], where the smartphone's built-in sensors are used to provide a driver safety scoring and tips on how to improve it.

Since this research wants to evaluate the use of a smartphone's built-in sensors for a setup that could be used in a V2V/V2I system, it will focus on the prediction of a vehicle's future trajectory, and compare the results with the use of more robust sensors mounted on a vehicle to predict the same future trajectory. Given that multiple sensors will be used, some type of sensor fusion will be needed to use the different measurements in the prediction.

3.2 Sensor Fusion Techniques

The MSDF techniques are used in many diverse fields, although most of the literature addresses the fields of military target tracking or autonomous robotics [21]. The MSDF is required to combine and process data, which has been traditionally performed by some form of Kalman [22] or Bayesian filters. Furthermore, there can be two ways of setting up an MSDF system: centralized or decentralized. While a centralized approach suffices for most common scenarios where the sensors are synchronous, a decentralized approach is more convenient when the sensors should be treated independently [23–28], as with asynchronous sensors.

In Reference 29, the authors discuss one solution they have developed: the OATFA, which evolved from their earlier research on an ASTF [30]. They base their technique in the IMM algorithm, but replaced the conventional Kalman filters with their OATFA (which contains several Kalman filters of its own). The OATFA treats each sensor separately, passing the output from each to a dedicated Kalman filter, departing from the idea that the best way to fuse data is to deliver it all to a central fusion engine. The results from the IMM–OATFA show position estimation errors half of those of what the conventional IMM produces. However, as pointed out by the same authors in Reference 31, all measurement data must be processed before the fusion algorithm is executed. With a similar approach as the technique described above, the authors of Reference 32 create asynchronous holds, where, from a sequence of inputs sampled at a slow sampling rate, it generates a continuous signal that may be discretized at a high sampling rate. Despite the benefits of making the asynchronous system into a synchronous one by using these methods, restrictions arise where, if for some reason, a sensor is delayed in providing its data or is offline for a few cycles. The whole system needs to wait, as it is designed to work with certain data at specific rates (Figure 3.5).

To evaluate whether smartphones can properly estimate future trajectories to be considered as an option to fill in the V2V/V2I implementation gap, a system to estimate a future position of a vehicle will be set up to determine where the vehicle will be 3 s later, which is based on the average human reaction time of 1.5 s to stop a vehicle [33]. Looking at 3 s ahead of time was chosen as a reference point that is double the reaction time of an average human being. In reality, this number will probably vary in relation to the speed and the type of the vehicle since a faster or heavier vehicle will need more time to slow down, but it is taken as a reference point.

3.3 Position Estimation with Kalman Filters

For this research, the core method chosen to estimate a future position of a vehicle is the use of KF. The KF [34] was first proposed in the 1960s and it has been the most commonly used technique in target tracking and robot navigation since. The basic KF has been presented as a form of Bayesian filter [35], which is an optimal estimator for linear Gaussian systems. From a series of noisy measurements, the KF is capable of estimating the state of the system in a two-step process: correct and then predict [36–38].

The elements of this state vector (x) are position, velocity, and acceleration of the vehicle. The position (x_v) and velocity (v_v) components of the state estimate have an x and y component to them (east-west and north-south directions), and the acceleration (a_v) has an n and t component to it (normal and tangential acceleration). So, the full state vector matrix will be $X = (x_x, x_y, v_x, v_y, a_n, a_t)$.

The estimated error covariance (P) for the state estimate is based on the relationships between each of the elements to the others. The error covariance matrix is a dataset that specifies the estimated accuracy in the observation errors between all possible pairs of vertical levels.

Together with P, the Jacobian matrix of the measurement model (H), the measurement noise covariance (R), and with the measurement noise (σ_m) are used to calculate the Kalman gain (K). Once the K is calculated, the system looks at the measured data (Z) to identify the error of the predicted position and uses it to adjust P.

The KF has a long history of accurately predicting future states of a moving target, and has been applied to many different fields [39–46], including transportation, which is why it was selected for this research. Because one KF estimates the future position of a vehicle using one spatial movement model, there is a need to set up several KFs to account for the different spatial states in which the vehicle can be found. Having multiple KF running in parallel at the same time requires a framework that can obtain one weighted answer.

This research opted for the use of IMM, which can calculate the probability of success of each KF model at every filter execution, providing a combined solution for the vehicle behavior [47–50]. These probabilities are calculated according to a Markov model for the transition between maneuver states, as detailed in Reference 51. To implement the Markov model, it is assumed that at each execution time, there is a probability p^{ij} that the vehicle will make the transition from model state i to state j.

In Johnson and Krishnamurthy's paper [52], they describe the IMM as a recursive suboptimal algorithm that consists of four core steps, interacting with the KF steps as illustrated in Figure 3.2.

The four-step IMM process starts with the calculation of the mixing probabilities, which uses the transition matrix and the previous iteration mode probabilities $\mu_{k-1}(i)$ to compute the normalized mixing probabilities

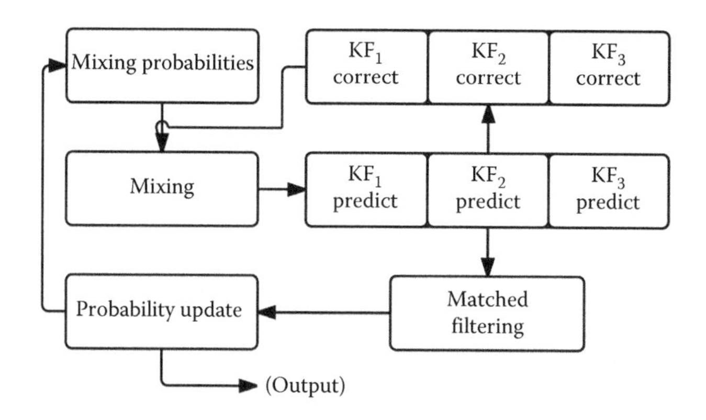

FIGURE 3.2
Flowchart for three KFs in an IMM framework.

$\mu_k(i|j)$. The mixing probabilities are recomputed each time the filter iterates before the mixing step.

The second step uses the mixing probabilities, which are used to compute new initial conditions for each of the n filters. The initial state vectors are formed as the weighted average of all of the filter state vectors from the previous iteration x_{k-1}^{oj}. The error covariance corresponding to each of the new state vectors is computed as the weighted average of the previous iteration's error covariance conditioned with the spread of the means $\left(P_{k-1}^{oj}\right)$.

The third step calculates mode-matched filtering; using the $\hat{x}_{k-1}^{0j}$ and P_{k-1}^{0j}, the bank of n Kalman filters produce outputs $\hat{x}_k^j$, the covariance matrix P_k^j, and the probability density function $f_n(z_k)$ for each filter (n).

The fourth set of calculations begins once the new initial conditions are computed; the filtering step generates a new state vector, error covariance, and likelihood function for each of the filter models. The probability update step then computes the individual filter probability ($\mu_k(j)$) as the normalized product of the likelihood function and the corresponding mixing probability normalization factor.

The estimate and covariance combination is used for output purposes only $\hat{x}_k = \sum_{j=1}^{n} \mu_k^j \cdot \hat{x}_k^j$; it is not part of the algorithm recursions.

3.4 Position Estimation Framework Using GPS and Accelerometer Sensors

For the evaluation of the use of smartphones to predict a future position of a vehicle, only those sensors common across all devices were selected, which, in this case are the GPS and accelerometer sensors. With the measurements obtained from these sensors, the future position estimation is obtained.

In this setup, the GPS sensor provides the location (s_x,s_y), the velocity (v), and the angle of direction (β) using north as the zero. Then the accelerometer provides normal acceleration (a_n) and tangential acceleration (a_t). The different models to be used in this setup are defined below, which represents the different spatial states the vehicle can be found in Equations 3.1 through 3.3:

Constant Location Model (CL)

$$\begin{aligned} s(k) &= s(k-1) + \sigma_{p_s} \\ v(k) &= 0 \\ a(k) &= 0 \end{aligned} \tag{3.1}$$

Constant Velocity Model (CV)

$$\begin{aligned} s(k) &= s(k-1) + v(k-1) \cdot \Delta k + \sigma_{p_s} \\ v(k) &= v(k-1) + \sigma_{p_v} \\ a(k) &= 0 \end{aligned} \tag{3.2}$$

Constant Acceleration Model (CA)

$$\begin{aligned} s(k) &= s(k-1) + v(k-1) \cdot \Delta k + \frac{1}{2} a(k-1) \cdot \Delta k^2 + \sigma_{p_s} \\ v(k) &= v(k-1) + a(k-1) \cdot \Delta k + \sigma_{p_v} \\ a(k) &= a(k-1) + \sigma_{p_a} \end{aligned} \tag{3.3}$$

The above three KF models are used as part of an IMM setup to merge each of the KF predictions and obtain one single predicted position, as described in Section 3.5. When not all sensor measurements are available to properly populate the three KF models, the DR approach described in Section 3.6 is used to handle the asynchronous data.

3.5 Multi-Sensor Data Fusion Setup

This research initially looks at predicting future positions by running the IMM system at 1 Hz with all vehicle-mounted (VM) sensors online in every iteration, and then running it at 10 Hz where some sensors are offline and measurements are missing in many of the iterations. A measurement could be missing, either due to the sensor not being able to take the measurement (system running at a faster frequency than the sensor, malfunction or no satellites in view for the GPS) or due to the processing CPU not being able to read/write fast enough. When a measurement is absent and the value is needed for the models, the missing

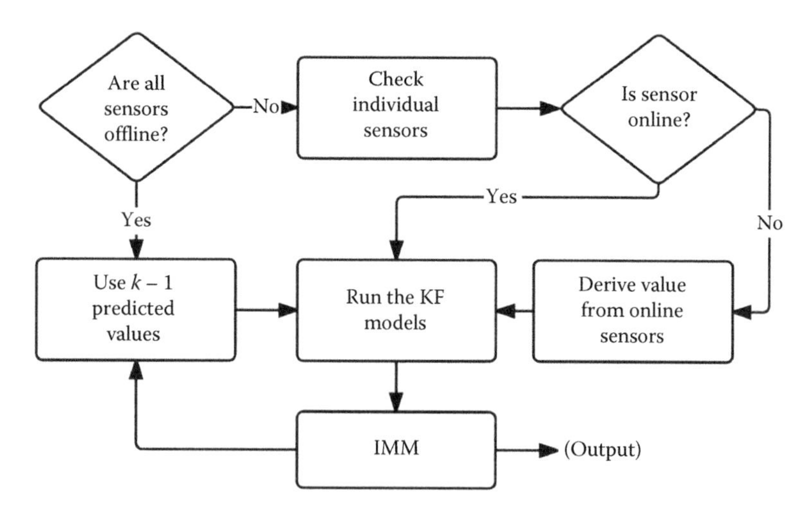

FIGURE 3.3
Flowchart of the position estimation framework used by the vehicle-mounted and smartphone sensors.

values are calculated from the measurements obtained by the remaining sensors based on previous real measurements, not estimations, when available.

In this research, because the system is running at the rate of the accelerometer (10 Hz), the missing measurements come from the GPS and ScanTool VM sensors, which run at only 1 Hz. Instead of using previously estimated values for the missing measurements, this setup uses the real measurements obtained from the accelerometer and uses the equations below to derive the missing measurements:

$$\begin{aligned} v(k) &= v(k-1) + a(k)\cdot\Delta k \\ s(k) &= s(k-1) + v(k-1)\cdot\Delta k + \frac{1}{2}a(k)\cdot\Delta k^2 \end{aligned} \tag{3.4}$$

Using this DR approach, as outlined in Figure 3.3, predictions are more accurate than waiting until all measurements are available again, or predicting these measurements a second time, using only previously estimated values. Only when all measurements are missing, which will be very unlikely, the system will use all the previously estimated values to feed the KF models and obtain the new position estimation.

3.6 Car and Smartphone Sensor Setup for a V2V/V2I System

As described toward the end of Section 3.2, this research will estimate the future position of the vehicle 3 s away using the framework described in

Section 3.4, which is something that could be shared with other vehicles (V2V) or with the infrastructure (V2I) as part of some collision avoidance system. As the communication standards are defined for the V2V and V2I systems, smartphones will fall into one of two different categories as briefly illustrated in Figure 3.4: (a) already supporting the technology required for the communication, or (b) needing an external device, illustrated as a black box, and connected via USB/Bluetooth/Wi-Fi to manage the communication aspect. This is briefly illustrated in Figure 3.4. Since smartphones and V2V/V2I are designed independently of each other, the use of an external device (black box) is more likely to be the case to enable smartphones to participate in a V2V/V2I system.

The VM sensors specifically set up for this specific task will be used, like manufacturers will implement in their vehicles. Smartphone (SP) sensors will also need to be evaluated when used for position estimation to determine if they yield similar results.

To properly evaluate if smartphones can be used in a V2V/V2I system, this research plans to set a baseline by running the VM measurements through the position estimation framework defined in Section 3.4, and calculate the position errors in the estimations by comparing them to the actual GPS data. Once the baseline is established and a determination of what are the amounts of prediction errors obtained, the individual SP measurements will be fed into the same position estimation framework and the error will be calculated in the position estimations. This research can then proceed to compare the results between the different sensors used and evaluate whether the smartphones' built-in sensors yield similar prediction errors or not.

The setup on the VM sensors consists of a Garmin 16HVS GPS receiver running at 1 Hz and a Crossbow 3-axis accelerometer running at 10 Hz. An AutoEnginuity ODBII ScanTool (which obtains the velocity from the vehicle's internal system at 1 Hz) is also available, but it will not be used in this evaluation because the smartphones this research is using do not have a way of connecting into the ODBII system. The data from the sensors used are postprocessed from the different log files recorded on an earlier date, and matched based on time stamps. Since these VM sensors were mounted on a van from the University of Connecticut, they will be labeled as UConn data throughout this research.

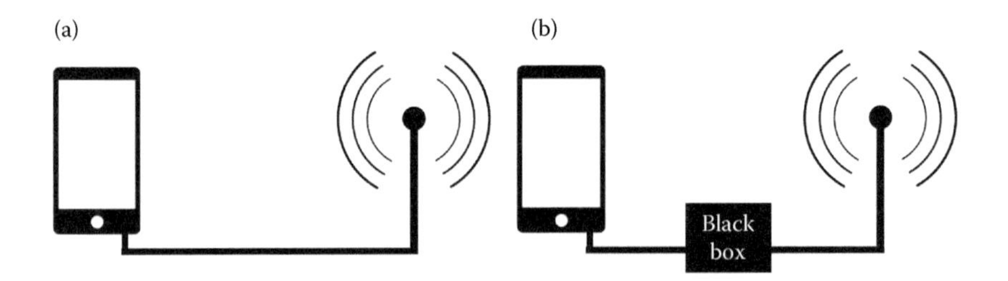

FIGURE 3.4
(a) SP supporting V2V/V2I communication technology and (b) SP needing external device (black box) for communication.

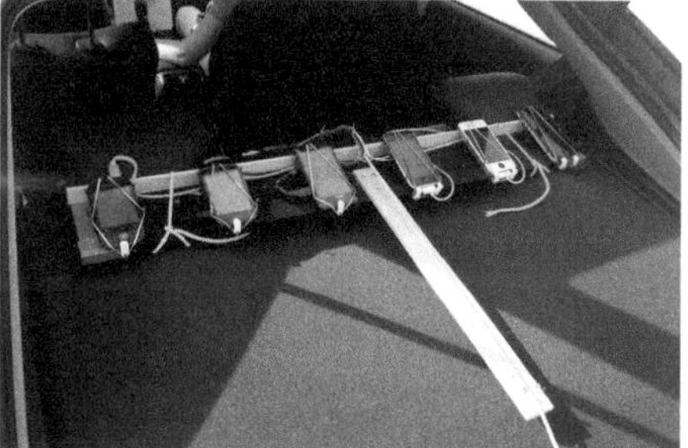

FIGURE 3.5
Smartphones securely mounted on the trunk of a hatchback vehicle.

For the smartphones used in this evaluation, some were selected from different manufacturers and at different price ranges, to identify if there is some limitation on which ones can accurately predict the future trajectory of a vehicle. Also, smartphones are used with different operating systems as well to improve the evaluation experiment and take that into account as well. They were mounted securely on the vehicle to ensure the accelerometer readings truly reflect the dynamics of the vehicle. Because several smartphones were running at the same time, they were mounted in the trunk where they would still have a clear view of the sky, as shown in Figure 3.5, but a more common implementation would be to mount only one of them on the dashboard. Figure 3.5 shows six smartphones, but one of them did not record any GPS data, so it had to be removed from this experiment. The smartphones used in the evaluation of accurately predicting future trajectories are listed in Table 3.1.

All smartphones listed above have a built-in accelerometer sensor that can take measurements at 10 Hz, but no details were found on their model or sensitivity. These smartphones also have a built-in GPS sensor, and only very basic information was found about them. iPhone 5 has an A-GPS/GLONASS sensor, while the other four smartphones do not have support for Global

TABLE 3.1
Smartphone Specs

Manufacturer	Model	OS	Release Date	Base Price
Alcatel	OneTouch 908F	Android 2.2	6/2011	$130
HTC	Desire C	Android 4.0	6/2012	$150
LG	Lucid VS 840	Android 2.3	4/2012	$300
Apple	iPhone 3GS	iOS 5.1	6/2009	$199[a]
Apple	iPhone 5	iOS 7.01	9/2012	$650

Note: Details about these smartphones obtained from http://www.gsmarena.com.

[a] Subsidized price; this model could not be purchased without a contract; the real price could be two or three times more.

Navigation Satellite System by the Russians (GLONASS). Also, both Apple smartphones can take measurements from the GPS sensor at 10 Hz, while the other three smartphones can only take measurements at 1 Hz.

Some smartphones also have a three-axis gyro sensor and a compass, which could be used as well to better estimate a position of a vehicle; but to match more closely the sensors mounted on the vehicle, and have a more equal comparison, they were not used in this experiment.

The measurements from the internal sensors of iPhone smartphones are recorded by running the SensorLog v1.4 application written by B. Thomas. The sensors' measurements on the Android smartphones are recorded using the Data Recording v1.0.2.0 application written by T. Wolf. The data used are labeled by smartphone manufacturer name, except when there are more than one device per manufacturer, in which case the data were labeled by model name.

To properly exercise the position estimation framework described in Section 3.4, the route shown in Figure 3.6a for this evaluation was selected, which contains several curves (smooth and sharp) and straight paths, driven at different speeds in the larger and smaller roads. There were also some traffic lights on the way, and even a U-turn, providing also some stop-and-go scenarios. The route driven, shown in Figure 3.6a, is approximately 44 km long and takes about 45 min to drive all of it.

It is important to keep in mind that, despite having all SP devices secured on the same area of the vehicle as shown in Figure 3.5, their sensors' sensitivity and tolerances are not the same, as visually shown in Figures 3.7 and 3.8.

Figure 3.7 shows the GPS readings of all SP and VM sensors, and, despite most of them being close to each other, they are not identical, and, therefore, will introduce some more error into the system when using the VM GPS as

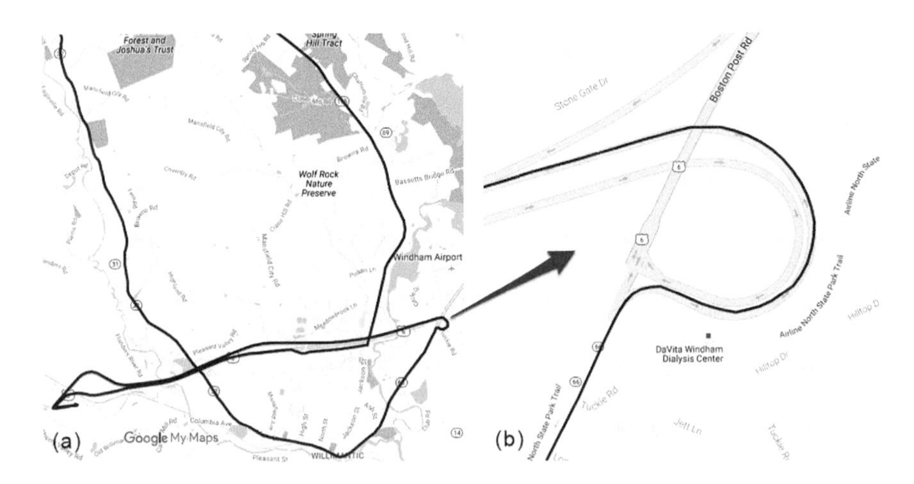

FIGURE 3.6
Map of the recorded route near the University of Connecticut: (a) whole trajectory and (b) selected curves.

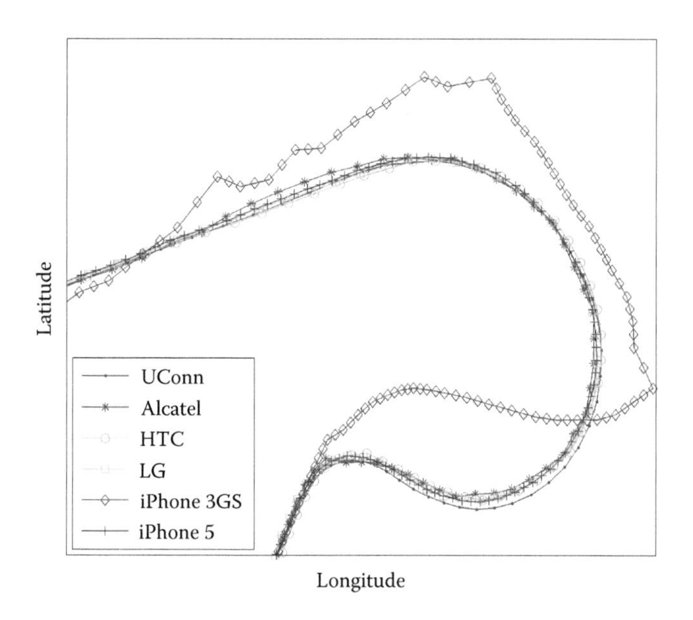

FIGURE 3.7
GPS measurements.

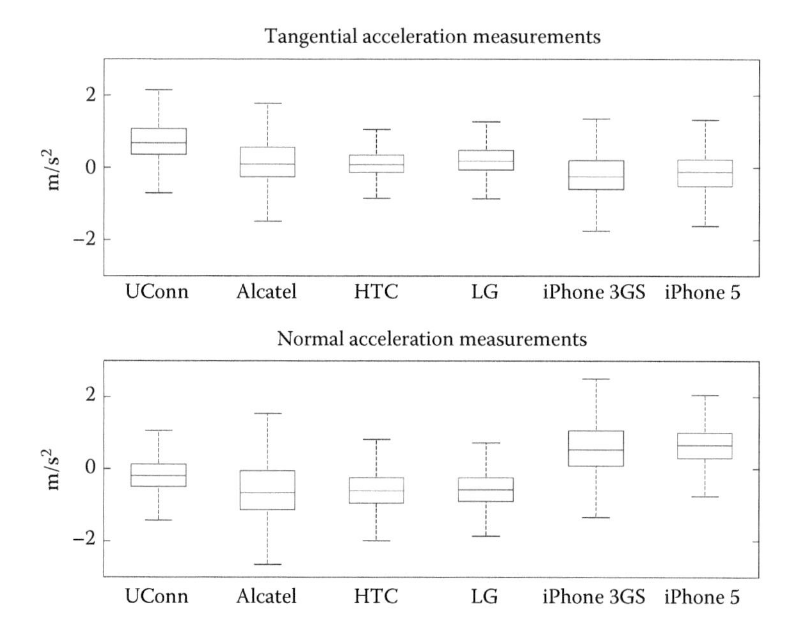

FIGURE 3.8
Accelerometer normal and tangential measurements.

a baseline. And, similarly, Figure 3.8 shows the accelerometer measurements for each of the devices used in this research, where it can be observed that some devices have a large variation of measurements compared to others.

3.7 Evaluation Criteria

To evaluate whether smartphones can properly predict a future trajectory and be considered as a possible solution to fill in the V2V/V2I implementation gap, the position estimation error between the VM sensors (UConn) and the SP sensors was selected using the same KF models and IMM framework. To start with, the position estimation error between both setups will be evaluated for the whole trajectory recorded. Also, since position estimation errors tend to increase during non-straight paths, this research will also divide the trajectory recorded into smooth and sharp curves. To determine whether a set of consecutive points in the trajectory is a curve or a straight line, the change in the heading (angle) between the current heading and the heading 2 s before was looked at; if more than 5°, then it was defined as a curve. And, to determine if the curve is a sharp one, the change has to be greater than 20°, otherwise it was defined as a smooth curve. For the selected curve shown in Figure 3.6b, the results of categorizing the trajectory based on heading is shown in Figure 3.9.

To calculate the position estimation error in each step, this research will compare the estimated position to the actual position measured by the GPS

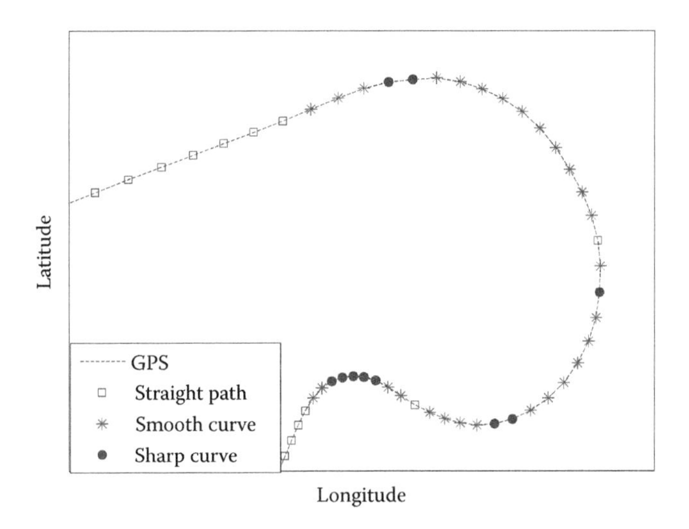

FIGURE 3.9
Classification of trajectory segments.

3 s later. This will allow a dataset of calculated errors to be built for the whole trajectory, which then can be divided into the route sections described in the previous paragraph. This research will look at average prediction errors and RMS prediction errors using Equation 3.5 to try to evaluate whether the sensors built into smartphones can properly fill in the V2V/V2I implementation gap:

$$\text{RMS} = \sqrt{\frac{\sum_{k}^{k'} (err_k - err_{k-1})^2}{(k' - k)}} \tag{3.5}$$

For this experiment, a tolerance of 10% from the position estimation errors obtained from the VM sensors will be used; therefore, if a smartphone yields more than 10% higher estimation error than the VM sensors, then it will be concluded that such a smartphone cannot properly predict a future trajectory, and hence could not be considered as a possible temporary solution to fill in the V2V/V2I implementation gap.

3.8 Experimental Evaluation

3.8.1 Dataset Characteristics

The characteristics of the complete recorded dataset are shown in Table 3.2, where the mean and standard deviation for the position difference between each second, velocity, normal acceleration, and tangential acceleration are displayed.

Looking at Table 3.2, it is quickly noticed that the values between the distance and velocity columns are very similar, as expected, because this research is measuring the position change every 1 s. Also, as mentioned in Section 3.6, the UConn data were obtained on an earlier date, so it can be seen that, overall, the University of Connecticut van was driven a little bit faster than the vehicle with the smartphones. Also, because all five smartphones were in the same vehicle, their sensor measurements should have been very similar, which is not the case in several places. For example, for sharp curves, the two iPhones seemed to be moving at a much higher speed than the other three devices, while during straight paths they seemed to be moving a little slower than the rest. The tangential acceleration for all devices seems to be fairly consistent across all devices, while the normal acceleration is not as consistent, especially when smooth and sharp curves were observed, which could imply that some sensors are more sensitive than others.

TABLE 3.2
Representative Datasets

	Device	Distance (m)	Velocity (m/s)	Normal Acceleration (m/s²)	Tangential Acceleration (m/s²)
	UConn	18.14 ± 8.99	17.95 ± 8.45	–0.17 ± 0.58	0.61 ± 0.60
Whole trajectory	Alcatel	16.31 ± 8.40	16.30 ± 7.95	–0.12 ± 0.73	0.66 ± 0.95
	HTC	16.21 ± 8.24	16.29 ± 8.01	–0.09 ± 0.56	0.64 ± 0.68
	LG	16.79 ± 9.13	16.45 ± 7.83	–0.18 ± 0.58	0.61 ± 0.64
	iPhone 3GS	16.59 ± 9.37	16.21 ± 7.96	–0.22 ± 0.74	0.51 ± 0.88
	iPhone 5	16.25 ± 9.09	16.24 ± 8.18	–0.11 ± 0.79	0.62 ± 0.82
	UConn	19.43 ± 8.29	19.02 ± 8.14	–0.16 ± 0.48	0.55 ± 0.57
Straight paths	Alcatel	18.76 ± 6.92	18.72 ± 6.70	–0.12 ± 0.72	0.64 ± 0.95
	HTC	18.40 ± 6.56	18.38 ± 6.51	–0.10 ± 0.54	0.62 ± 0.57
	LG	18.98 ± 7.74	18.42 ± 6.65	–0.18 ± 0.56	0.59 ± 0.50
	iPhone 3GS	16.96 ± 9.04	16.63 ± 7.77	–0.20 ± 0.74	0.50 ± 0.88
	iPhone 5	16.23 ± 9.24	16.23 ± 8.29	–0.10 ± 0.78	0.60 ± 0.73
	UConn	16.19 ± 9.26	15.81 ± 8.03	–0.29 ± 0.89	0.74 ± 0.63
Smooth curves	Alcatel	14.66 ± 7.90	14.58 ± 7.36	–0.11 ± 0.75	0.73 ± 0.98
	HTC	12.76 ± 8.32	12.92 ± 7.99	–0.07 ± 0.64	0.78 ± 0.91
	LG	14.01 ± 9.28	13.53 ± 7.65	–0.21 ± 0.64	0.68 ± 0.82
	iPhone 3GS	12.47 ± 11.31	11.99 ± 8.49	–0.34 ± 0.70	0.58 ± 0.87
	iPhone 5	17.99 ± 6.64	17.86 ± 5.83	–0.16 ± 0.89	0.82 ± 1.29
Sharp curves	UConn	8.52 ± 9.17	9.91 ± 7.65	–0.10 ± 0.89	1.01 ± 0.68
	Alcatel	7.13 ± 9.60	7.49 ± 8.35	–0.16 ± 0.79	0.63 ± 0.88
	HTC	5.29 ± 9.05	5.97 ± 8.68	–0.07 ± 0.64	0.59 ± 0.94
	LG	5.48 ± 8.86	7.09 ± 8.01	–0.16 ± 0.66	0.59 ± 1.03
	iPhone 3GS	16.30 ± 12.13	13.50 ± 9.49	–0.15 ± 1.05	0.70 ± 0.89
	iPhone 5	12.00 ± 8.54	12.18 ± 8.54	–0.50 ± 1.04	0.62 ± 1.58

Note: Values represent median ± standard deviation of all sensor measurements collected by each device (~25,000 data points).

3.8.2 Position Estimation Setup

To set up the IMM, it is necessary to calculate the transition probability matrix, so the GPS position measurements for the whole trajectory shown in Figure 3.6a were used. From this full GPS log that contained multiple scenarios, it was determined which transition occurs frame by frame by comparing the actual measurements from the GPS to the smoothed measurements. The smoothing of the data was done with a rolling window using a combination of median smoothing, split the sequence, and Hann's sequence, which removed any abrupt changes from the data. The type of spatial change, such as no change, a constant change, and so on, determines each transition. Similarly, by calculating the covariance of the differences in the measurements to each other, the measurement noise covariance matrix (R) was obtained. And last, by calculating the covariance of the differences in the measurements compared to their respective x and y components, the process covariance noise (Q) for each KF was obtained. From this type of information, calculating the frequency the vehicle changes from one state to another, the transition probability matrix is derived.

Next, each of the devices was run through the same IMM system using the KF models described in Section 3.4, and for each new measurement obtained from any of the sensors, the system predicts the position of where the vehicle is going to be 3 s later in time.

3.8.3 Evaluation of Position Estimation Error by System Rate

This part of the experiment is to make sure the system defined in Section 3.5, which runs at the rate of its fastest sensor, yields better results than running the system at the rate of its slowest sensor. First, the VM sensor's dataset (UConn) is run though the IMM system at the rate of 1 Hz when all sensors are online at each iteration. Then, the same dataset is run through the same IMM system but running at 10 Hz, so now missing values from offline sensors are calculated based on measurements from online sensors using a DR approach. For this part of the experiment, all of the VM sensors available were used, including the AutoEnginuity ODBII ScanTool.

Accuracy in predicting where the vehicle will be 3 s later in time is the factor to observe to be able to evaluate if running the IMM system at the rate of its fastest sensor yields smaller errors. Calculating the RMS prediction error by comparing each predicted position with each actual GPS measurement 3 s late in time, Table 3.3 is compiled.

Taking a quick look at Table 3.3, it can be observed that the RMS prediction errors for the system running at 10 Hz are smaller than when the system is run at 1 Hz, especially during curves. Even during straight paths, some improvements can be observed because the system is detecting a change in the vehicle's speed faster, and can accordingly recalculate its prediction.

TABLE 3.3

RMS Prediction Error (3 s Ahead)

	UConn at 1 Hz	UConn at 10 Hz
Whole trajectory	1.68 ± 3.21	1.29 ± 2.64
Straight paths	1.27 ± 2.66	0.92 ± 1.65
Smooth curves	2.60 ± 3.73	2.28 ± 4.87
Sharp curves	3.43 ± 4.98	2.51 ± 4.36

Note: Contains values representing median prediction error in meters ± standard deviation.

3.8.4 Evaluation of Position Estimation Error by Device

Now that it is shown that running this research's system at the rate at the fastest sensor yields better predictions of future positions, all smartphones are also run through this setup and their corresponding prediction errors are recorded in Table 3.4. This table displays the RMS distance between the predicted and actual positions. This prediction error can only be calculated when the time stamps between the predicted position and GPS reading match. It is assumed that the GPS reading is correct and it calculates the distance vector to the predicted position. Then the mean and standard deviation was calculated of all the calculated RMS error vectors for the whole trajectory and also for the defined segment types.

As expected, the prediction error was less during straight paths, and it increased during curves. Based on the values recorded in Table 3.4, the prediction errors can double during curves. Also, the prediction errors for smooth curves were better than during the sharp curves, which makes sense because, in a smooth curve, the vehicle is changing its heading less abruptly than in a sharp curve, allowing the system more time to recalculate and correct its next prediction.

This research also observed that the prediction error was not the same between all devices, and sometimes a device that has a small prediction error in one segment type may not be as good as that on a different segment type, making it hard to draw conclusions from Table 3.4. In spite of these results, if one looks at the percent deviation of prediction errors compared to the UConn results, it can be narrowed down to the HTC and LG cellphones having the smaller prediction errors overall and meeting the tolerance of no more than 10% more error than obtained with the UConn sensors.

Figure 3.10 is another way of representing the prediction errors for each of the devices in the different segment types. The boxplots display the median value as the solid line dividing the box into two, and then the upper and lower half of the boxes represent the interquartiles, which together represent 50% of the calculated prediction errors. The upper whisker indicates that 75% of the errors fall below it, and the lower whisker indicates the 25%

TABLE 3.4

RMS Prediction Error (3 s Ahead)

	UConn	Alcatel	HTC	LG	iPhone 3GS	iPhone 5
Whole trajectory	1.29 ± 2.64	1.88 ± 2.62	1.00 ± 1.03	1.13 ± 1.25	2.34 ± 3.00	1.41 ± 2.08
Straight paths	0.92 ± 1.65	1.77 ± 2.56	0.89 ± 1.10	0.98 ± 1.13	2.21 ± 2.81	1.35 ± 2.05
Smooth curves	2.28 ± 4.87	1.89 ± 2.14	0.97 ± 0.97	1.07 ± 1.23	3.49 ± 4.02	1.86 ± 2.40
Sharp curves	2.51 ± 4.36	2.11 ± 3.83	1.21 ± 1.23	1.43 ± 1.59	4.78 ± 4.70	2.13 ± 2.05

Note: Values representing median prediction error in meters ± standard deviation by all devices for trajectory shown in Figure 3.6 (~25,000 data points).

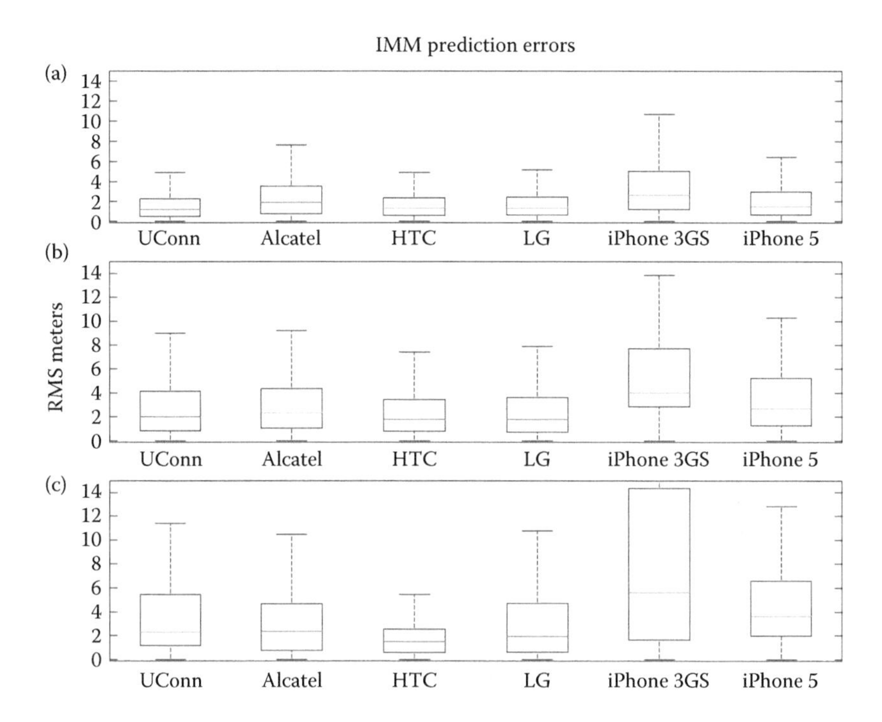

FIGURE 3.10
Prediction errors during (a) straight paths, (b) smooth curves, and (c) sharp curves.

marker. With this in mind, it can be seen that for the straight paths, except for the iPhone 3GS, the boxes are very short, which means that the prediction errors have a high level of agreement. One can also see small boxes in the smooth and sharp curves for the HTC and LG, so it can be observed that their predictions are fairly consistent most of the time, unlike the boxplot for the iPhone 3GS where it is a very large box indicating a very low level of agreement between the predictions. Also, the lower the boxes to the x-axis, the smaller the prediction errors; so a small box close to the x-axis, like the HTC in sharp curves, or the UConn in straight lines, represents a very accurate prediction system. Again, looking at the boxplots for the five smartphones, one can visually pick the HTC and LG to be fairly good, then maybe Alcatel and iPhone 5, though it looks like iPhone 5 is not as reliable as Alcatel during sharp curves.

When looking at iPhone 3GS results, both in Table 3.4 and Figure 3.10, it can be observed that this device has prediction errors much larger than other devices. It seems this device has a problem, losing its signal quite often, introducing more errors to what was assumed to be the "true" position. Looking more into this topic, we have found several Apple discussion forums (https://discussions.apple.com) where users have reported very inaccurate GPS locations when using iPhone 3GS running iOS 5.

TABLE 3.5

RMS Prediction Error (3 s Ahead)

	UConn	Alcatel	HTC	LG	iPhone 3GS	iPhone 5
Whole trajectory	1.46 ± 0.97	3.22 ± 3.64	1.81 ± 1.87	1.56 ± 1.65	6.34 ± 6.95	1.79 ± 2.18
Straight paths	0.69 ± 0.53	2.96 ± 4.60	1.42 ± 2.11	0.77 ± 0.58	4.50 ± 5.76	0.95 ± 0.74
Smooth curves	1.69 ± 0.54	3.13 ± 2.18	1.75 ± 1.43	1.84 ± 2.13	11.57 ± 6792	1.57 ± 0.85
Sharp curves	1.98 ± 0.97	3.52 ± 3.81	2.67 ± 2.03	2.19 ± 1.26	14.73 ± 6.24	2.69 ± 3.12

Note: Values represent median prediction error in RMS meters ± standard deviation by all devices for the selected curve (~550 data points).

To look at a subset of the whole route shown in Figure 3.6a, a couple of curves were selected as shown in Figure 3.6b, and the results represented in a similar way, but only for this small subset of the dataset. This selected segment of the route represents only 0.8 km (36% straight path, 44% smooth curve, and 20% sharp curve), which takes around 10 s to go through.

When looking at Table 3.5, the first difference that might be observed when comparing it to the results for the whole route shown in Table 3.4 is that the average prediction error for the whole trajectory of the selected subset is different. In this case, straight paths are a small percentage of the whole selected subset while smooth curves are the most abundant. For this very specific set of curves, the UConn data are better than any of the smartphones in all trajectory types. The LG device yields the smallest prediction errors of all the smartphones, and still within the selected 10% tolerance when compared to the UConn results. The next best devices seem to be HTC and iPhone 5 smartphones, where, despite having prediction errors over the 10% tolerance, their prediction errors are around 20% worse than the UConn results.

Another difference one can observe in Table 3.5 is that, unlike the results in Table 3.3, the HTC device did not seem to perform as well in this selected set of curves than when evaluated over the whole route. Even when looking at the results for smooth and sharp curves, the HTC results were always worse than the UConn prediction errors, which is not the case when looking at the data in Table 3.4. In Table 3.5, it seems that most smartphones performed worse than the UConn setup in this set of curves. Since this is consistent across all smartphones, it can be concluded that there was something on the curves that affected the prediction, like bumps or maybe unleveled pavement, especially over the sharp curve section.

A visual representation of the prediction errors during the small subset of curves previously mentioned is shown in Figure 3.11, where each predicted trajectory is compared to the actual GPS position measured 3 s later. One can observe that for some devices there is a smooth trajectory of predicted positions, like for the UConn and LG, closely followed by the HTC, but it can also be observed that some other devices are constantly correcting its predicted position drastically, causing all those spikes during the curves.

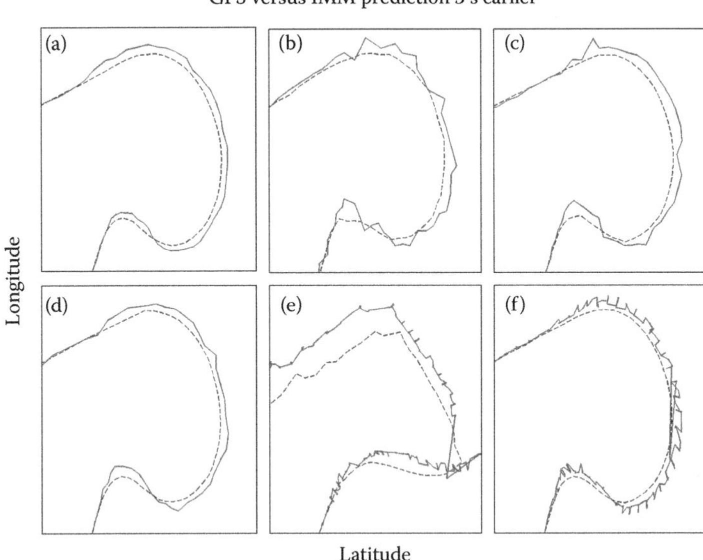

FIGURE 3.11
Dotted blue lines represent GPS measurements while solid red lines represent IMM position predicted 3 s earlier for (a) UConn, (b) Alcatel, (c) HTC, (d) LG, (e) iPhone 3GS, and (f) iPhone 5.

One positive thing of looking at the predicted position errors as shown in Figure 3.11 is that, despite the conclusions obtained from Table 3.5 that iPhone 5 was predicting a future position better than several of the other devices, this research would not think this is a reliable device after looking at Figure 3.11. Therefore, even though Figure 3.11 cannot be used by itself to draw some final conclusions, it is a very useful addition to Table 3.5.

An important lesson learned in this research was that the accelerometer sensor could provide misleading information based on road slopes. Table 3.6 shows a data subset where velocity was increasing but tangential acceleration was decreasing, implying that the vehicle was slowing down. Looking

TABLE 3.6
Misleading Accelerometer Readings

Time Stamp	Altitude (m)	Velocity (m/s)	Tangential Acceleration (m/s^2)
33:17	117.93	18.06	–0.508
33:18	118.03	18.30	–0.593
33:19	118.12	18.53	–1.061
33:20	118.27	18.77	–1.168
33:21	118.80	19.07	–1.858

Note: Misleading accelerometer measurements due to altitude changes.

at the data, it can be observed that the vehicle was actually speeding up, and the decrease on the tangential acceleration was actually because the road was going up (altitude increase) and not because the vehicle was slowing down.

3.9 Conclusions

The built-in sensors of some smartphones were evaluated to predict the future trajectory of a vehicle and their prediction errors compared to those obtained by using VM sensors. Results shown in Tables 3.4 and 3.5 indicate that smartphones yield similar or better prediction errors and could, therefore, be used in older vehicles to participate in a V2V/V2I system. Though smartphone price seemed to play a small role, the HTC smartphone is one of the cheaper ones used in this experiment, and it performed quite well in some scenarios. The more expensive LG device yielded more reliable results in more, so price could be a factor, but then the iPhone 5, being the most expensive one, did not contribute well to the price factor. If the results of the trajectory prediction errors for the whole trajectory driven for this experiment (shown in Table 3.4) are observed, it can be concluded that all smartphones, except for iPhone 3GS, yield prediction errors up to 10% more than the UConn errors. Actually, it can be observed that the results from the smartphones perform, in most scenarios described in Table 3.4, even better than the baseline set by the VM sensors (UConn), some even almost 50% better. With these results, it can be safely concluded that the trajectory prediction accuracy of the smartphones' sensors is sufficiently reliable to be considered for a temporary hook to enable an older vehicle to participate in a V2V/V2I system.

Future research will include the use of other sensors found in some of the smartphones used (gyroscope and magnetic compass) to evaluate if they improve prediction of a vehicle's trajectory and how these predicted trajectories can fit into collision avoidance systems.

References

1. V. Di Lecce, A. Amato, Route planning and user interface for an advanced intelligent transport system, *IET Intelligent Transport Systems* 5(3), 2011: 149–158, doi: 10.1049/iet-its.2009.0100.
2. X. Xu, T. Xia, A. Venkatachalam, D. Huston, The development of a high speed ultrawideband ground penetrating radar for rebar detection, *Journal of Engineering Mechanics* 139(3), 2013: 272–285, doi: 10.1061/(ASCE)EM.1943-7889.0000458.

3. T. Taleb, A. Benslimane, K. Ben Letaief, Toward an effective risk-conscious and collaborative vehicular collision avoidance system, *IEEE Transactions on Vehicle Technology* 59(3), 2010: 1474–1486, doi: 10.1109/TVT.2010.2040639.
4. S. Lefèvre, D. Vasquez, C. Laugier, A survey on motion prediction and risk assessment for intelligent vehicles, *ROBOMECH Journal* 1(1), 2014: 1–14, doi: 10.1186/s40648-014-0001-z.
5. F. Jimenez, J. Eugenio Naranjo, Improving the obstacle detection and identification algorithms of a laserscanner-based collision avoidance system, *Transportation Research Part C: Emerging Technologies* 19(4), 2011: 658–672, http://dx.doi.org/10.1016/j.trc.2010.11.001.
6. R. Toledo-Moreo, M.A. Zamora-Izquierdo, Collision avoidance support in roads with lateral and longitudinal maneuver prediction by fusing GPS/IMU and digital maps, *Transportation Research Part C: Emerging Technologies* 18(4), 2010: 611–625, doi: 10.1016/j.trc.2010.01.001.
7. C. Hakyoung, L. Ojeda, J. Borenstein, Accurate mobile robot dead-reckoning with a precision-calibrated fiber-optic gyroscope, *IEEE Transactions on Robotic Automation* 17(1), 2001: 80–84, doi: 10.1109/70.917085.
8. R. Toledo-Moreo, M. Pinzolas-Prado, J. Manuel Cano-Izquierdo, Maneuver prediction for road vehicles based on a neuro-fuzzy architecture with a low-cost navigation unit, *IEEE Transactions on Intelligent Transportation Systems* 11(2), 2010: 498–504, doi: 10.1109/TITS.2009.2039011.
9. J. Ueki, J. Mori, Y. Nakamura, Y. Horii, H. Okada, Development of vehicular-collision avoidance support system by inter-vehicle communications, *IEEE 59th Vehicle Technology Conference*, Vol. 5, pp. 2940–2945, Milan, Italy, May 17–19, 2004.
10. Y. Wang, S. Wang, M. Tan, Path generation of autonomous approach to a moving ship for unmanned vehicles, *IEEE Transactions on Industrial Electronics* 62(9), 2015: 5619–5629, doi: 10.1109/TIE.2015.2405904.
11. J.J. Blum, A. Eskandarian, A reliable link-layer protocol for robust and scalable intervehicle communications, *IEEE Transactions on Intelligent Transportation Systems* 8(1), 2007: 4–13, doi: 10.1109/TITS.2006.889441.
12. Jin Wen-Long, W. Recker, An analytical model of multihop connectivity of inter-vehicle communication systems, *IEEE Transactions on Wireless Communications* 9(1), 2010: 106–112, doi: 10.1109/TWC.2010.01.05545.
13. B. Dalla Chiara, F. Deflorio, S. Diwan, Assessing the effects of inter-vehicle communication systems on road safety, *IET Intelligent Transport Systems* 3(2), 2009: 225–235, doi: 10.1049/iet-its:20080059.
14. S. Sohaib, D.K.C. So, Asynchronous cooperative relaying for vehicle-to-vehicle communications, *IEEE Transactions on Communications* 61(5), 2013: 1732–1738, doi: 10.1109/TCOMM.2013.031213.120008.
15. M. Fogue, P. Garrido, F.J. Martinez, J.C. Cano, C.T. Calafate, P. Manzoni, Automatic accident detection: Assistance through communication technologies and vehicles, *IEEE Transactions on Vehicular Technology Magazine* 7(3), 2012: 90–100, doi: 10.1109/MVT.2012.2203877.
16. Liu Chunhua, K.T. Chau, Wu Diyun, Gao Shuang, Opportunities and challenges of vehicle-to-home, vehicle-to-vehicle, and vehicle-to-grid technologies, *Proceedings of the IEEE* 101(11), 2013: 2409–2427, doi: 10.1109/JPROC.2013.2271951.
17. J.C. Ferreira, V. Monteiro, J.L. Afonso, Vehicle-to-anything application (V2Anything App) for electric vehicles, *IEEE Transactions on Industrial Informatics* 10(3), 2014: 1927–1937, doi: 10.1109/TII.2013.2291321.

18. D. Freidman, V2V: Cars Communicating to Prevent Crashes, Deaths, Injuries. U.S. DOT, February 3, 2014, http://www.dot.gov/fastlane/v2v-cars-communicating-prevent-crashes-deaths-injuries
19. J. Gorselanv, Cars That Can Last for 250,000 Miles (Or More), Forbes Corp., March 14, 2013, http://www.forbes.com/sites/jimgorzelany/2013/03/14/cars-that-can-last-for-250000-miles/
20. Cambridge Mobile Telematics, Data Analytics and Processing, August 2013, http://www.cmtelematics.com/product/data-analytics
21. R.C. Luo, C.C. Yih, K.L. Su, Multisensor fusion and integration: Approaches, applications, and future research directions, *IEEE Transactions on Sensors* 2, 2002: 107–119, doi: 10.1109/JSEN.2002.1000251.
22. J.B. Gao, C.J. Harris, Some remarks on Kalman filters for the multisensor fusion, *Elsevier Information Fusion* 3, 2002: 191–201, doi: 10.1016/S1566-2535(02)00070-2.
23. S.C. Felter, An overview of decentralized Kalman filter techniques, *Proceedings of IEEE of the 1990 Southern Tier Technical Conference*, pp. 79–87, Binghampton, New York, April 25, 1990.
24. M. Hua, T. Bailey, P. Thompson et al., Decentralized solutions to the cooperative multi-platform navigation problem, *IEEE Transactions on Aerospace and Electronic Systems* 47(2), 2011: 1433–1449, doi: 10.1109/TAES.2011.5751268.
25. E.M. Nebot, M. Bozorg, H.F. Durrant-Whyte, Decentralized architecture for asynchronous sensors, *Autonomous Robots* 6(2), 1999: 147–164, doi: 10.1023/A:1008883411036.
26. M. Vallee, M. Merdan, W. Lepuschitz, G. Koppensteiner, Decentralized reconfiguration of a flexible transportation system, *IEEE Transactions on Industrial Informatics* 7(3), 2011: 505–516, doi: 10.1109/TII.2011.2158839.
27. H.M. Wang, Q. Yin, X. Xia, Fast Kalman equalization for time-frequency asynchronous cooperative relay networks with distributed space-time codes, *IEEE Transactions on Vehicular Technology* 59(9), 2010: 4651–4658, doi: 10.1109/TVT.2010.2076352.
28. G.A. Watson, T.R. Rice, A.T. Alouani, An IMM architecture for track fusion, *Signal Processing, Sensor Fusion, and Target Recognition (Proc. SPIE 4052)* IX, 2000: 2–13.
29. A.T. Alouani, T.R. Rice, On optimal asynchronous track fusion, *Proceedings of 1st IEEE Australian Symposium on Data Fusion*, Adelaide, Australia, November 21–22, 1996.
30. G.A. Watson, T.R. Rice, A.T. Alouani, Optimal track fusion with feedback for multiple asynchronous measurements, *Proceedings of SPIE Acquisition, Tracking and Pointing XIV*, Orlando, Florida, July 7, 2000.
31. M.W. Owen, S.C. Stubberud, Interacting multiple model tracking using a neural extended Kalman filter, *International Joint Conference on Neural Networks*, Vol. 4, pp. 2788–2791, Washington DC, July 10–16, 1999.
32. J. Wang, S.Y. Chao, A.M. Agogiono, Validation and fusion of longitudinal positioning sensors in AVCS, *Proceedings of 1999 American Control Conference*, Vol. 3, pp. 2178–2182, San Diego, California, June 2–4, 1999.
33. M. Green, How long does it take to stop? Methodological analysis of driver perception-brake times, *Transportation Human Factors* 2(3), 2000: 195–216, doi: 10.1207/STHF0203_1.
34. R. Kalman, A new approach to linear filtering and prediction problems, *Transactions on ASME* 82, 1960: 34–45, doi: 10.1115/1.3662552.

35. Y. Ho, R. Lee, A Bayesian approach to problems in stochastic estimation and control, *IEEE Transactions on Automatic Control* 9(4), 1964: 333–339, doi: 10.1109/TAC.1964.1105763.
36. G. Welch, G. Bishop, An introduction to the Kalman filter, SIGGRAPH Course Notes, 2001.
37. Y. Bar-Shalom, X.R. Li, T. Kirubarajan, *Estimation with Applications to Tracking and Navigation*, Wiley and Sons, Hoboken, New Jersey, 2001.
38. Y. Bar-Shalom, X.R. Li, *Estimation and Tracking: Principles, Techniques and Software*, Artech House, Norwood, Massachusetts, 1993.
39. S.Y. Chen, Kalman filter for robot vision: A survey, *IEEE Transactions on Industrial Electronics* 9(11), 2012: 4409–4420, doi: 10.1109/TIE.2011.2162714.
40. Chih-Lung Lin, Yi-Ming Chang, Chia-Che Hung, Chun-Da Tu, Cheng-Yan Chuang, Position estimation and smooth tracking with a fuzzy-logic-based adaptive strong tracking Kalman filter for capacitive touch panels, *IEEE Transactions on Industrial Electronics* 62(8), 2015: 5097–5108, doi: 10.1109/TIE.2015.2396874.
41. M. Ho Kim, S. Lee, K.C. Lee, Kalman predictive redundancy system for fault tolerance of safety-critical systems, *IEEE Transactions on Industrial Informatics* 6(1), 2010: 46–53, doi: 10.1109/TII.2009.2020566.
42. X. Xu, Z. Xiong, X. Sheng, J. Wu, X. Zhu, A new time synchronization method for reducing quantization error accumulation over real-time networks: Theory and experiments, *IEEE Transactions on Industrial Informatics* 9(3), 2013: 1659–1669, doi: 10.1109/TII.2013.2238547.
43. B. Aubert, J. Régnier, S. Caux, D. Alejo, Kalman-filter-based indicator for online interturn short circuits detection in permanent-magnet synchronous generators, *IEEE Transactions on Industrial Electronics* 62(3), 2015: 1921–1930, doi: 10.1109/TIE.2014.2348934.
44. R.K. Singleton, E.G. Strangas, S. Aviyente, Extended Kalman filtering for remaining-useful-life estimation of bearings, *IEEE Transactions on Industrial Electronics* 62(3), 2015: 1781–1790, doi: 10.1109/TIE.2014.2348934.
45. C.F. Graetzel, B.J. Nelson, S.N. Fry, A dynamic region-of-interest vision tracking system applied to the real-time wing kinematic analysis of tethered Drosophila, *IEEE Transactions on Automation Science and Engineering* 7(3), 2010: 463–473, doi: 10.1109/TASE.2009.2031094.
46. F. Auger, M. Hilairet, J.M. Guerrero, E. Monmasson, T. Orlowska-Kowalska, S. Katsura, Industrial applications of the Kalman filter: A review, *IEEE Transactions on Industrial Electronics* 60(12), 2013: 5458–5471, doi: 10.1109/TIE.2012.2236994.
47. L. Hong, Multirate interacting multiple model filtering for target tracking using multirate models, *IEEE Transactions on Automation Control* 44(7), 1999: 1326–1340, doi: 10.1109/9.774106.
48. E. Mazor, Interacting multiple model methods in target tracking: A survey, *IEEE Transactions on Aerospace Electronic Systems* 34(1), 1998: 103–123, doi: 10.1109/7.640267.
49. L. Wenling, J. Yingmin, Location of mobile station with maneuvers using an IMM-based cubature Kalman filter, *IEEE Transactions on Industrial Electronics* 59(11), 2012: 4338–4348, doi: 10.1109/TIE.2011.2180270.
50. S.J. Lee, Y. Motai, H. Choi, Tracking human motion with multichannel interacting multiple model, *IEEE Transactions on Industrial Informatics* 9(3), 2013: 1751–1763, doi: 10.1109/TII.2013.2257804.

51. Y. Bar-Shalom, H.A.P. Blom, The interacting multiple model algorithm for systems with Markovian switching coefficients, *IEEE Transactions on Automation Control* 33(8), 1988: 780–783, doi: 10.1109/9.1299.
52. L.A. Johnson, V. Krishnamurthy, An improvement to the interactive multiple model (IMM) algorithm, *IEEE Transactions on Signal Processing* 49(12), 2001: 2909–2923, doi: 10.1109/78.969500.

4

Summary of Vehicle Trajectories' Prediction Methods Evaluated

This research focused on the prediction of vehicle trajectory that could be used as part of a V2V/V2I collision avoidance system. The mathematical estimation algorithm selected was the KF, using multiple prediction models in a multiple models framework. The KF models were created based on the possible spatial states a vehicle could be found in, such as constant location, constant velocity, constant acceleration, and constant jerk.

Prediction improvements contributed by this research are as follows:

- Elimination of incorrect predictions when the future location is found outside of an actual road, with the assumption that the vehicle would remain on the road while traveling
- Define a framework where the predictions could be obtained and corrected at the rate of its fastest sensor instead of only when all sensors are online
- Improve the trajectory predictions by increasing the process noise covariance when it is not based on a real sensor measurement (dynamic Q)

Table 4.1 lists prediction methods used in this research and displays them in order of worse to better. The IMM was expected to outperform MMAE, and in our research, it displayed improvements of 20%. The DRWDE method showed that a more frequent recalculation of the prediction improved the prediction errors by 25% over the IMM. And the IMM+GIS, presented in Chapter 1, clearly outperformed even the DRWDE by 20%.

The last part of the research, as explained in Chapter 3, focused on looking for an alternative to vehicle-mounted sensors to predict a vehicle's trajectory to enable older vehicles to participate in a modern V2V/V2I collision avoidance system. For this purpose, this research evaluated the use of a smartphone's built-in sensors to predict a vehicle's trajectory.

As shown in Table 4.2, for a given selected curve with a sharp and smooth curve, the built-in sensors of at least three of the smartphones used yielded

TABLE 4.1

Vehicle Trajectories' Prediction Improvements

Prediction Method	Prediction Error
MMAE (1 Hz)	Worst method because it can only choose one KF model at a time
IMM (1 Hz)	Better method because it is a weighted average of KF models
DRWDE (10 Hz)	Noticeable improvements over traditional IMM implementation
IMM+GIS (1 Hz)	Lower prediction errors than all the other methods during curves

Note: Error in predicting a vehicle trajectory for the selected curve, from largest to smallest.

TABLE 4.2

Summary of Vehicle Trajectories' Prediction Error Using Smartphones

Device	Prediction Error
iPhone 3GS	Very large errors due to weak GPS signal locks, which introduced a lot of errors
Alcatel	Large errors, which make this device not very reliable for predicting a trajectory
HTC	Errors around 22% larger than UConn baseline
iPhone 5	Errors around 16% larger than UConn baseline
LG	Errors around 11% larger than UConn baseline
UConn	Vehicle-mounted sensors used as a baseline to compare the other devices to this one

Note: Error in predicting a vehicle trajectory for the selected curve, from largest to smallest.

similar results as the UConn vehicle-mounted sensors when used for predicting a vehicle trajectory. This confirms that it may be possible to use some smartphones as a temporary solution to connect older vehicles into a modern V2V/V2I collision avoidance system until all vehicles have the appropriate built-in technology.

Appendix

A.1 Acronym Definitions

Acronym	Definition
ASTF	Asynchronous/synchronous track fusion
CA	Constant acceleration
CJ	Constant jerk
CL	Constant location
CPU	Central processing unit
CV	Constant velocity
USDOT	United States Department of Transportation
DR	Dead-reckoning
DRWDE	Dead-reckoning with dynamic error
FAA	Federal Aviation Administration
GIS	Geographic information system
GPS	Global positioning system
IMM	Interacting multiple model
ITS	Intelligent transportation system
KF	Kalman filter
MMAE	Multiple models adaptive estimation
MSDF	Multi-sensor data fusion
OATFA	Optimal asynchronous track fusion algorithm
RMS	Root mean square
SP	Smartphone
TMA	Target motion analysis
V2I	Vehicle-to-infrastructure
V2V	Vehicle-to-vehicle
VM	Vehicle-mounted
WAAS	Wide area augmentation system

A.2 Symbol Definitions

Symbol	Definition
P	Estimated error covariance
H	Jacobian of the measurement model
A	Jacobian of the system model with respect to state
K	Kalman gain
X	State vector
Z	Measurement data
s_v	Position component (x/y) of the state vector
v_v	Velocity component (x/y) of the state vector
a_v	Acceleration component (x/y) of the state vector
a_n	Normal acceleration
a_t	Tangential acceleration
σ_m	Measurement noise
R	Measurement noise covariance
N	Number of filters
σ_p	Prediction noise
λ	Probability
Q	Process noise covariance
x	State estimate
β	Angle of movement direction (bearing)
err	Error between actual and estimated positions
Dk	Time interval
p^{ij}	Transition probability matrix

A.3 Microsoft Visual Basic 6 Code Used in Chapter 1

Visual Basic Module	Description
Main_Module.bas	Starting point where system setup variables are defined and KF/IMM/MMAE are called from
Map_Module.bas	Contains functions used to visually display GPS and estimated location points on the map
Calc_offset_2.bas	Contains functions used to determine/correct if the estimated position falls outside of a road
GPS_Module.bas	Contains functions to read GPS log, stores it in variables, and displays the location on the map
imm4.bas	Contains functions to execute the KF and IMM steps to estimate the future location

The **Main_Module.bas** file is the starting point of this code. In it, the user defines the different settings required to run the system, and then the code calls the appropriate KF and IMM/MMAE frameworks to estimate the future position of the vehicle. The estimated position is then displayed on the map during runtime, so the user can visually see the system working step by step.

```
Attribute VB_Name = "Main_Module"
Global APP_PATH, fs1, fout_res, KFMM, SNAP2ROAD, SHOWARC,
SHOWLINE, CALCOFFSET, SYSTEMLOOP, IMM_STEPS, REC_NUM,
STOP_NUM
Global sec_ahead, DISPLAY_ESTIMATION

Sub Main()

    EMULATE_GPS = True    'True if emulating from file or
False if using GPS

    APP_PATH = "Z:\dropbox\UVM\PhD\CAS3\
kalman_filter(VB)"

    LOGNAME = "newgps_essexjct_123_w.txt"

    KFMM = "IMM4"   ' Options are MMAE or IMM4 or EKF

    CALCOFFSET = False
    SHOWLINE = False        'Never used other than
visually inspect distance for CALCOFFSET

    SNAP2ROAD = True        'GIS error correction
    SHOWARC = False         'Never used other than to
visually inspect the arc used to SNAP2ROAD

    SYSTEMLOOP = True       'Loop system another 2 times
to get 3sec estimation

    IMM_STEPS = True       'Run EKF with or without IMM.

    sec_ahead = 3           'number of seconds to
calculate/display data (for display in label3 only)
    REC_NUM = 102   '102      'record number to start
keeping track of results (for display in label3 only)
    STOP_NUM = 123   '1000     'record number to stop
keeping track of results (for display in label3 only)

    DISPLAY_ESTIMATION = False  'display estimated
location on map
```

```
    ZOOM_ACTUAL = 1
    ZOOM_AUTO = "True"

    ZOOM_IN_BUFFER = 0
    ZOOM_OUT_BUFFER = 6
    ZOOM_CNT = 0
    ZOOM_DESIRED = 0
    ZOOM_TMP1 = 0
    ZOOM_TMP2 = 0

    ZOOM_0_SPEED = 35      '35 mph or 60 kph
    ZOOM_1_SPEED = 45      '45 mph or 80 kph
    ZOOM_2_SPEED = 60      '55 mph or 100 kph
    ZOOM_3_SPEED = 70      '65 mph or 120 kph

    ZOOM_4_DIST = 3        '5 miles    or 10 km
    ZOOM_3_DIST = 1        '3 miles    or 5 km
    ZOOM_2_DIST = 0.5      '1 miles    or 2 km
    ZOOM_1_DIST = 0.25     '0.5 miles  or 1 km
    ZOOM_0_DIST = 0.1      '0.25 miles or 0.5 km

    ZOOM_0_ALTITUDE = 0
    ZOOM_1_ALTITUDE = 4
    ZOOM_2_ALTITUDE = 8
    ZOOM_3_ALTITUDE = 13
    ZOOM_4_ALTITUDE = 20

    highlight_changed = False
    HIGHLIGHT_DIST = ZOOM_4_DIST

    GPS_TIMER_INTERVAL = 1   'Value is in miliseconds
(1000=1sec)

    ARRIVED_AT_TURN = 0.2

    MDIForm1.Show

End Sub

Public Sub close_all()
      frmGPS.Timer1.Enabled = False
      Unload frmGPS
      Unload MDIForm1
End Sub
```

The **Map_module.bas** file contains the functions used to display different data points on the map (MS MapPoint) and projected trajectories, to visually inspect how the system is reacting to the vehicle's spatial changes.

```
Attribute VB_Name = "Map_Module"

Function Update_Current_Location()
    'Delete previous location Pin
    Set objPin = objMap.FindPushpin("Current Location")
    objPin.Delete

    'Insert current location pin
    Set objLoc = objMap.GetLocation(curlat, curlong)
    objMap.AddPushpin objLoc, "Current Location"

    'Change current location to GREEN
    objMap.DataSets.Item(1).Symbol = 30

    'Center on current location
    objLoc.Location.GoTo

strLine = ""

Locx_prev2 = Locx_prev1 & " "
Locx_prev1 = Locx & " "
Locx = Convert_Deg2Rad(curlong)

Locy_prev2 = Locy_prev1
Locy_prev1 = Locy
Locy = Convert_Deg2Rad(curlat)

'VELOCITY = VELOCITY / 3600  'converting from mph to mps

Vx_prev2 = Vx_prev1
Vx_prev1 = Vx
'Vx = VELOCITY * Sin(ANGLE)
Vx = Locx - Locx_prev1

Vy_prev2 = Vy_prev1
Vy_prev1 = Vy
'Vy = VELOCITY * Cos(ANGLE)
Vy = Locy - Locy_prev1

Ax_prev1 = Vx_prev1 - Vx_prev2
Ax = Vx - Vx_prev1

Ay_prev1 = Vy_prev1 - Vy_prev2
Ay = Vy - Vy_prev1

Validate_Algorithm1
Validate_Algorithm2
Validate_Algorithm3
```

```
Write_results

End Function

Sub Validate_Algorithm1()
    'Algorithm1 - Projected path based on change of
location only (assumes constant velocity)

    next1lat = Locx + (Locx - Locx_prev1)
    next1long = Locy + (Locy - Locy_prev1)

    next2lat = next1lat + (Locx - Locx_prev1)
    next2long = next1long + (Locy - Locy_prev1)

    next3lat = next2lat + (Locx - Locx_prev1)
    next3long = next2long + (Locy - Locy_prev1)

    secv_x1 = next1lat
    secv_x2 = next2lat
    secv_x3 = next3lat
    secv_y1 = next1long
    secv_y2 = next2long
    secv_y3 = next3long

End Sub

Sub Validate_Algorithm2()
    'Algorithm2 - Projected path based on change of
velocity (assumes constant acceleration)

    next1vx = Vx + (Vx - Vx_prev1)
    next1vy = Vy + (Vy - Vy_prev1)

    next1lat = Locx + next1vx
    next1long = Locy + next1vy

    next2vx = next1vx + (Vx - Vx_prev1)
    next2vy = next1vy + (Vy - Vy_prev1)

    next2lat = next1lat + next2vx
    next2long = next1long + next2vy

    next3vx = next2vx + (Vx - Vx_prev1)
    next3vy = next2vy + (Vy - Vy_prev1)

    next3lat = next2lat + next3vx
    next3long = next2long + next3vy

    seca_x1 = next1lat
    seca_x2 = next2lat
```

```
    seca_x3 = next3lat
    seca_y1 = next1long
    seca_y2 = next2long
    seca_y3 = next3long

End Sub

Sub Validate_Algorithm3()
    'Algorithm3 - Projected path based on change of
acceleration

    next1ax = Ax + (Ax - Ax_prev1)
    next1ay = Ay + (Ay - Ay_prev1)

    next1vx = Vx + next1ax
    next1vy = Vy + next1ay

    next1lat = Locx + next1vx
    next1long = Locy + next1vy

    next2ax = next1ax + (Ax - Ax_prev1)
    next2ay = next1ay + (Ay - Ay_prev1)

    next2vx = next1vx + next2ax
    next2vy = next1vy + next2ay

    next2lat = next1lat + next2vx
    next2long = next1long + next2vy

    next3ax = next2ax + (Ax - Ax_prev1)
    next3ay = next2ay + (Ay - Ay_prev1)

    next3vx = next2vx + next3ax
    next3vy = next2vy + next3ay

    next3lat = next2lat + next3vx
    next3long = next2long + next3vy

    secj_x1 = next1lat
    secj_x2 = next2lat
    secj_x3 = next3lat
    secj_y1 = next1long
    secj_y2 = next2long
    secj_y3 = next3long

End Sub

Sub Write_results()
   fout_res.WriteLine (strLine)
```

```
End Sub

Function Convert_Dist2Deg(dist, dir)
    If (dir = "LAT") Then
        'Dist is in miles, dir in LAT
        '1" = 0.0192136389 miles approx
        Convert_Dist2Deg = (dist / 0.0192136389) / 3600
'Converting miles into seconds and then into degrees
    Else
        'Dist is in miles, dir in LONG
        '1" = 0.0192107222 miles approx
        Convert_Dist2Deg = (dist / 0.0192107222) / 3600
'Converting miles into seconds and then into degrees
    End If
End function
```

The **Calc_offset_2.bas** file contains the functions used to determine if the estimated position falls outside of an actual road, and corrects it until it is over a road (GIS).

```
Attribute VB_Name = "Calc_offset_2"

'Module that snaps GPS location to actual nearest road

Sub Start(longitude, latitude, prev_lon, prev_lat)
   Dim streets

    curlong = longitude
    curlat = latitude

    tst = GetStreetAddr(latitude, longitude, prev_lat,
prev_lon, streets)

End Sub

Function FindAddr(lat, lon, streets)
   Dim Item()
   Dim Loc As Location
   Dim StreetResults As MapPoint.FindResults
   Dim objResult As Object
   Dim icnt As Integer

   Set Loc = objMap.GetLocation(lat, lon)
   gpsLocx = objMap.LocationToX(Loc)
```

```
   gpsLocy = objMap.LocationToY(Loc)
   Set StreetResults = objMap.ObjectsFromPoint(gpsLocx, 
gpsLocy)

   If (StreetResults.ResultsQuality < geoNoGoodResult) 
Then
    icnt = 0
    For Each objResult In StreetResults
        tmp = objResult.name
        If (InStr(1, tmp, ",") > 0) Then
            FindAddr = icnt + 1
            curlat = lat
            curlong = lon
        End If
      icnt = icnt + 1
      Next
   Else
      FindAddr = 0
   End If

End Function

Function GetStreetAddr(lat, lon, prev_lat, prev_lon, 
streets)
   Dim angle As Integer

   streets = ""
   distance = 0
   Direction = ""
   streets_cnt = FindAddr(lat, lon, streets)
   If (streets_cnt = 0) Then
      tmp = CalcInCircle(lat, lon, prev_lat, prev_lon, 
streets, distance, angle)
      streets_cnt = 1
   End If
   GetStreetAddr = streets_cnt

End Function

Function CalcXY(centerLat, centerLon, angle, radius, lat, 
lon)

   'angle in radians
   'radius in degrees
   'centerLat/Lon in degrees
   a = radius * Math.Sin(angle)
   B = radius * Math.Cos(angle) '/ Math.Cos(Kalman_
Filters.Convert_Deg2Rad(centerLat)) 'correct longitude 
for circle
```

```
    lat = centerLat + a
    lon = centerLon + B
    CalcXY = 0

End Function

Function CalcInCircle(centerLat, centerLon, prev_lat,
prev_lon, streets, distance, angle)

     CalcInCircle = False

     FDistanceResolution = 2       'x feet gap
     FMaxCount = 100               'x times
FDistanceResolution both ways

     Dy = Kalman_Filters.Convert_Deg2Rad(centerLat - prev_
lat) 'in radians
     Dx = Kalman_Filters.Convert_Deg2Rad(centerLon - prev_
lon) 'in radians
     radius = Math.Sqr(Dx * Dx + Dy * Dy)  'in radians

     Dxft = 16033650.94 * Dx 'in feet
     Dyft = 20904250.56 * Dy 'in feet
     radiusft = Math.Sqr(Dxft * Dxft + Dyft * Dyft) 'in
feet

     Dydeg = (centerLat - prev_lat) 'in degrees
     Dxdeg = (centerLon - prev_lon) 'in degrees
     radiusdeg = Math.Sqr(Dxdeg * Dxdeg + Dydeg * Dydeg)
'in feet

     angle_0 = Math.Atn(Dydeg / Dxdeg) '=arctan(Dy/Dx) in
     radians
     angle_0 = Abs(angle_0)

     'determining quadrant and adjusting angle
     If Dx > 0 And Dy < 0 Then
         'QUAD = 3 from 180 to 270 degrees
         'Simulating QUAD 4 for opposite x direction from
270 to 360 degrees
         angle_0 = (2 * 3.14) - angle_0
     ElseIf Dx > 0 And Dy > 0 Then
         'QUAD = 2 from 90 to 180 degrees
         'Simulating QUAD 1 for opposite x direction from
0 to 90 degrees
         angle_0 = angle_0
     ElseIf Dx < 0 And Dy < 0 Then
         'QUAD = 4 from 270 to 360 degrees
```

```
        'Simulating QUAD 3 for opposite x direction from
180 to 270 degrees
        angle_0 = (3.14) + angle_0
    Else
        'QUAD = 1 from 0 to 90 degrees
        'Simulating QUAD 2 for opposite x direction from
90 to 180 degrees
        angle_0 = (3.14) - angle_0
    End If

    angle_0_deg = Kalman_Filters.Convert_Rad2Deg(angle_0)
    angle_0_rad = angle_0 'in degrees

'   Circum = Round(radius * 2 * 3.14, 0)
    Circum = radiusft * 2 * 3.14 'in feet
    Count = Circum / FDistanceResolution
'    AngleInc = 360 / Count  'in degrees
    AngleInc = (2 * 3.14) / Count   'in radians

If Form1.Check1 Then
    aaa = 1
End If
        Loop_count2 = 1
        While (Loop_count2 <= FMaxCount And CalcInCircle
= False)

           'Checking clock wise direction
           angle_cw = angle_0_rad + AngleInc *
Loop_count2
           tmp = CalcXY(prev_lat, prev_lon, angle_cw,
radiusdeg, lat, lon)
If SHOWARC Then
    Set objPin = objMap.FindPushpin("TTT_CW")
    objPin.Delete
    Set objLoc = objMap.GetLocation(lat, lon)
    objMap.AddPushpin objLoc, "TTT_CW"
End If
           tmp = FindAddr(lat, lon, streets)
           If (tmp > 0 And CalcInCircle = False) Then
    Set objPin = objMap.FindPushpin("TTT_CW")
    objPin.Delete
    Set objLoc = objMap.GetLocation(lat, lon)
    objMap.AddPushpin objLoc, "TTT_CW"
              curlong = lon 'in degrees
              curlat = lat  'in degrees
              CalcInCircle = True
           End If

           'Checking counter clock wise direction
```

```
            angle_ccw = angle_0_rad - AngleInc *
Loop_count2
            tmp = CalcXY(prev_lat, prev_lon, angle_ccw,
radiusdeg, lat, lon)
If SHOWARC Then
    Set objPin = objMap.FindPushpin("TTT_CCW")
    objPin.Delete
    Set objLoc = objMap.GetLocation(lat, lon)
    objMap.AddPushpin objLoc, "TTT_CCW"
End If
            tmp = FindAddr(lat, lon, streets)
            If (tmp > 0 And CalcInCircle = False) Then
    Set objPin = objMap.FindPushpin("TTT_CCW")
    objPin.Delete
    Set objLoc = objMap.GetLocation(lat, lon)
    objMap.AddPushpin objLoc, "TTT_CCW"
                curlong = lon 'in degrees
                curlat = lat  'in degrees
                CalcInCircle = True
            End If

            Loop_count2 = Loop_count2 + 1
        Wend

 End function
```

The **GPS_Module.bas** file contains the code that reads the GPS log, stores its measurements in variables, and then displays the measurement on the map using the MS MapPoint software.

```
Attribute VB_Name = "GPS_Module"
Private Declare Function BitBlt Lib "gdi32" (ByVal
hDestDC As Long, _
    yVal X As Long, ByVal Y As Long, ByVal nWidth As
Long,_
    ByVal nHeight As Long, ByVal hSrcDC As Long, ByVal
xSrc As Long, _
    ByVal ySrc As Long, ByVal dwRop As Long) As Long

Dim allsats(12) As Integer
Dim elev(12) As Integer
Dim azim(12) As Integer
Dim sigstr(12) As Integer

Dim fs As Object
Dim maxspd
```

```
Global EMULATE_GPS, LOGNAME, angle, VELOCITY
Global FOut As Object
Global flag
Global strUnparsed As String
Global curlat, curlong', prevlat_tmp, prevlong_tmp,
prevlat, prevlong
Global curSpeed, GPS_STATUS
Global GPS_TIMER_INTERVAL
Global strCurrent As String
Global section_no As String

Sub Start_GPS_Tracking()

If (EMULATE_GPS = True) Then
   Set fs = CreateObject("Scripting.FileSystemObject")
   Set FOut = fs.OpenTextFile(APP_PATH & "\" & LOGNAME, 1)'
   open for reading
Else

    If frmGPS.MSComm2.PortOpen = False Then
        frmGPS.MSComm2.CommPort = 6
        frmGPS.MSComm2.Settings = "4800,N,8,1"
        frmGPS.MSComm2.PortOpen = True

        'Turn on GPMRC msgs every 2 seconds
          frmGPS.MSComm2.Output = "$PRWIILOG,RMC,A,T,2,0"
& vbCrLf
        'Turn on GPGGA msgs every 2 seconds
          frmGPS.MSComm2.Output = "$PRWIILOG,GGA,A,T,2,0"
& vbCrLf
        'Turn off PRWIZCH messages
          frmGPS.MSComm2.Output = "$PRWIILOG,ZCH,V,,," &
vbCrLf
     End If
End If

End Sub

Sub Stop_GPS_Tracking()
    If frmGPS.MSComm2.PortOpen = True Then
        frmGPS.MSComm2.PortOpen = False
    End If
End Sub

Sub ParseInput()

    If Len(strUnparsed) = 0 Then Exit Sub
    crpos = InStr(1, strUnparsed, vbCr)
```

```
    Do While crpos > 0
       strCurrent = Left(strUnparsed, crpos - 1)
       strUnparsed = Mid(strUnparsed, crpos + 2)
       crpos = InStr(1, strUnparsed, vbCr)

       If (EMULATE_GPS = False) Then
           'Make unit start sending raw NMEA data
            If InStr(1, strCurrent, "ASTRAL") > 0 Then
                 frmGPS.MSComm2.Output = "ASTRAL" & vbCr
                'Turn on GPMRC msgs every 2 seconds
                  frmGPS.MSComm2.Output =
"$PRWIILOG,RMC,A,T,2,0" & vbCrLf
                'Turn on GPMRC msgs every 2 seconds
                  frmGPS.MSComm2.Output =
"$PRWIILOG,GSA,A,T,5,0" & vbCrLf
                'Turn on GPMRC msgs every 2 seconds
                  frmGPS.MSComm2.Output =
"$PRWIILOG,GSV,A,T,6,0" & vbCrLf
                'Turn on GPGGA msgs every 2 seconds
                  frmGPS.MSComm2.Output =
"$PRWIILOG,GGA,A,T,4,0" & vbCrLf
                'Turn off PRWIZCH messages
                  frmGPS.MSComm2.Output =
"$PRWIILOG,ZCH,V,,," & vbCrLf
                'Preset lat/long/time
                  frmGPS.MSComm2.Output = "$PRWIINIT,V,,,
3338.6000,N,11155.0000,W,0,0,M,0,T,190500,021000" & vbCrLf
            End If
       End If

       If InStr(1, strCurrent, "###") Then
          section_no = Mid(strCurrent, 5)
       End If

       If InStr(1, strCurrent, "$GPRMC") Then ParseRMC
strCurrent
       If InStr(1, strCurrent, "$GPGSV") Then ParseGSV
strCurrent
       If InStr(1, strCurrent, "$GPGGA") Then ParseGGA
strCurrent

       frmGPS.txtDisplay.Text = frmGPS.txtDisplay.Text &
strCurrent & vbCrLf
       If Len(frmGPS.txtDisplay.Text) > 1500 Then frmGPS.
txtDisplay.Text = Right(frmGPS.txtDisplay.Text, 1500)
       frmGPS.txtDisplay.SelStart = Len(frmGPS.
txtDisplay.Text)

    Loop
```

```
End Sub

Sub ParseRMC(inrmc As String)
    '$GPRMC,031736,V,4043.3101,N,07317.5308,W,0.000,0.0,1
20800,14.1,W*52
    Dim rmcdata(12) As String

    flag = 0
    token = ","
    tokenpos = 0
    oldtokenpos = 1
    crpos = InStr(1, inrmc, vbCr)
    outrmc = inrmc
    For n = 1 To 11
        tokenpos = InStr(oldtokenpos + 1, outrmc, token)
        curstr = Mid(outrmc, oldtokenpos + 1, tokenpos -
oldtokenpos - 1)
        If n = 2 Then
            If Len(curstr) > 0 Then frmGPS.lblUTC =
"Time: " & Left(curstr, 2) & ":" & Mid(curstr, 3, 2) &
":" & Right(curstr, 2) & " GMT"
        End If

        If n = 3 Then

        If (Left(curstr, 1) = "V") Then
            frmGPS.lblStatus.ForeColor = RGB(255, 0, 0)
            frmGPS.lblStatus.Caption = "AWAITING FIX"
            GPS_STATUS = "YELLOW"
        ElseIf (Left(curstr, 1) = "A") Then
            frmGPS.lblStatus.ForeColor = RGB(0, 128, 0)
            frmGPS.lblStatus.Caption = "SATS OK"
            GPS_STATUS = "GREEN"
        End If
        End If

        If n = 4 Then
            frmGPS.lblLat = "Lat: " & StrToDeg(curstr)
            curlat = StrToDeg(curstr)
        End If
        If n = 5 Then frmGPS.lblLat = frmGPS.lblLat &
curstr
        If n = 6 Then
            frmGPS.lblLong = "Long: " & -StrToDeg(curstr)
            curlong = -StrToDeg(curstr)
        End If
        If n = 7 Then frmGPS.lblLong = frmGPS.lblLong &
curstr
```

```
        If n = 8 Then
VELOCITY = curstr
        If (curstr * 1.151) > maxspd Then maxspd =
(curstr * 1.151)
            frmGPS.lblSpeed = "Speed (mph): " &
Format(curstr * 1.151, "0.00") & " (" & Format(maxspd,
"0.00") & " MAX)"
            DrawCurrSpeed frmGPS.picSpeed, (curstr * 1.151)
            curSpeed = curstr * 1.151
        End If

        If n = 9 Then
If (Len(curstr) = 0) Then
curstr = "0.00"
End If
        frmGPS.lblTrack = "Track: " & curstr
        PlotBearings frmGPS.picBearings, 0# + curstr
        End If

        If n = 10 Then frmGPS.lblDate = "Date: " &
Left(curstr, 2) & "-" & Mid(curstr, 3, 2) & "-" &
Right(curstr, 2)
        oldtokenpos = tokenpos
    Next
    flag = 1
    ' lblLat, lblLong
    '   **  1) UTC Time
    '   **  2) Status, V = Navigation receiver warning
    '   **  3) Latitude
    '   **  4) N or S
    '   **  5) Longitude
    '   **  6) E or W
    '   **  7) Speed over ground, knots
    '   **  8) Track made good, degrees true
    '   **  9) Date, ddmmyy
    '   ** 10) Magnetic Variation, degrees
    '   ** 11) E or W
    '   ** 12) Checksum

    If (GPS_STATUS = "GREEN") Then
        Kalman_Filters.Update_Current_Location
    End If

End Sub

Sub ParseGGA(ingga As String)
' $GPGGA,120757,5152.985,N,00205.733,W,1,06,2.5,121.9,M,4
9.4,M,,*52
```

```
Dim ggadata(12) As String

token = ","
tokenpos = 0
oldtokenpos = 1
crpos = InStr(1, ingga, vbCr)
outgga = ingga

For n = 1 To 11
    tokenpos = InStr(oldtokenpos + 1, outgga, token)
    curstr = Mid(outgga, oldtokenpos + 1, tokenpos -
oldtokenpos - 1)
    If n = 2 Then
    If Len(curstr) > 0 Then frmGPS.lblUTC = "Time: " &
Left(curstr, 2) & ":" & Mid(curstr, 3, 2) & ":" &
Right(curstr, 2) & " GMT"
    End If
    If n = 3 Then
        frmGPS.lblLat = "Lat: " & StrToDeg(curstr)
        curlat = StrToDeg(curstr)
    End If
    If n = 4 Then frmGPS.lblLat = frmGPS.lblLat & curstr

    If n = 5 Then
        frmGPS.lblLong = "Long: " & StrToDeg(curstr)
        curlong = -StrToDeg(curstr)
    End If
    If n = 6 Then
    frmGPS.lblLong = frmGPS.lblLong & curstr
    End If

    If n = 10 Then
    altft = (curstr / 39.36) * 12
    frmGPS.lblAlt = "Alt: " & Format(altft, "0.00") &
" ft"
    ScrollImg frmGPS.picAlt
    DrawCurrAlt frmGPS.picAlt, 0 + altft, 200, 100
    End If

    oldtokenpos = tokenpos
Next

'1 time of fix (hhmmss),
'2 latitude,
'3 N/S,
'4 longitude,
'5 E/W,
'6 Fix quality (0=invalid, 1=GPS fix, 2=DGPS fix),
```

```
'7 number of satellites being tracked,
'8 horizontal dilution of position,
'9 altitude above sea level,
'10 M (meters),
'11 height of geoid (mean sea level) above WGS84 ellipsoid,
'12 time in seconds since last DGPS update,
'13 DGPS station ID number,
'14 checksum

End Sub

Sub ParseGSV(ingsv As String)
'$GPGSV,2,2,08,05,31,055,29,11,14,290,,15,13,221,28,23,13
,152,*7B
On Error Resume Next

token = ","
tokenpos = 0
oldtokenpos = 1
crpos = InStr(1, ingsv, vbCr)
outgsv = ingsv
endsent = 0

' GSV token
    tokenpos = InStr(oldtokenpos + 1, outgsv, token)
    curstr = Mid(outgsv, oldtokenpos + 1, tokenpos -
oldtokenpos - 1)
    oldtokenpos = tokenpos

' Tot GSV msgs in block
    tokenpos = InStr(oldtokenpos + 1, outgsv, token)
    curstr = Mid(outgsv, oldtokenpos + 1, tokenpos -
oldtokenpos - 1)
    oldtokenpos = tokenpos

' Cur GSV msg num
    tokenpos = InStr(oldtokenpos + 1, outgsv, token)
    curstr = Mid(outgsv, oldtokenpos + 1, tokenpos -
oldtokenpos - 1)
    oldtokenpos = tokenpos
    If Int(curstr) = 1 Then
        frmGPS.lstSats.Clear
        frmGPS.lstSats.AddItem "Sat " & vbTab & "Elev " &
vbTab & "Azi " & vbTab & "Str"
    End If

' Sats in view
    tokenpos = InStr(oldtokenpos + 1, outgsv, token)
```

```
    curstr = Mid(outgsv, oldtokenpos + 1, tokenpos -
oldtokenpos - 1)
    oldtokenpos = tokenpos
    satsinview = Int(curstr)

' Up to 4 sats
For n = 1 To 4
    tokenpos = InStr(oldtokenpos + 1, outgsv, token)
    curstr = Mid(outgsv, oldtokenpos + 1, tokenpos -
oldtokenpos - 1)
    satname = Int(curstr)
    oldtokenpos = tokenpos

    tokenpos = InStr(oldtokenpos + 1, outgsv, token)
    curstr = Mid(outgsv, oldtokenpos + 1, tokenpos -
oldtokenpos - 1)
    satelev = curstr
    If Len(curstr) = 0 Then satelev = "??"
    oldtokenpos = tokenpos

    tokenpos = InStr(oldtokenpos + 1, outgsv, token)
    curstr = Mid(outgsv, oldtokenpos + 1, tokenpos -
oldtokenpos - 1)
    satazim = curstr
    If Len(curstr) = 0 Then satazim = "??"
    oldtokenpos = tokenpos

    tokenpos = InStr(oldtokenpos + 1, outgsv, token)
    If tokenpos = 0 Then
        satstrg = "??"
        endsent = 1
    Else
        curstr = Mid(outgsv, oldtokenpos + 1, tokenpos -
oldtokenpos - 1)
        satstrg = curstr
        If Len(curstr) = 0 Then satstrg = "??"
        oldtokenpos = tokenpos
    End If

    frmGPS.lstSats.AddItem satname & vbTab & satelev &
vbTab & satazim & vbTab & satstrg

    PlotSat 0 + satname, 0 + satelev, 0 + satazim, 0 +
satstrg
    If endsent = 1 Then Exit Sub
Next

'End If

End Sub
```

```
Sub PlotSat(satname As Integer, satelev As Integer,
satazim As Integer, satstrg As Integer)

Dim X, Y
PI = 3.14159265358979

' Correct so that N = up
satazim = (satazim - 90) Mod 360

X = 180 + satelev * Cos((satazim * PI / 180))
Y = 180 + satelev * Sin((satazim * PI / 180))
frmGPS.picCompass.ForeColor = RGB(255, 0, 0)
frmGPS.picCompass.PSet (X, Y), RGB(255, 0, 0)

End Sub

Sub PlotBearings(Pic As Object, satbear As Double)
Dim X, Y
PI = 3.14159265358979

' Correct so that N = up
angle = (satbear * PI / 180)
satbear = (satbear - 90) Mod 360
X = 180 + 100 * Cos((satbear * PI / 180))
Y = 180 + 100 * Sin((satbear * PI / 180))
Pic.Cls
PlotLines Pic

Pic.Line (180, 180)-(X, Y)

End Sub

Sub PlotLines(Pic As Object)

Pic.Line (0, 0)-(360, 360), RGB(192, 192, 192)
Pic.Line (0, 180)-(360, 180), RGB(192, 192, 192)
Pic.Line (180, 0)-(180, 360), RGB(192, 192, 192)
Pic.Line (360, 0)-(0, 360), RGB(192, 192, 192)
Pic.Circle (180, 180), 45, RGB(192, 192, 192)
Pic.Circle (180, 180), 90, RGB(192, 192, 192)
Pic.Circle (180, 180), 135, RGB(192, 192, 192)
Pic.Circle (180, 180), 180, RGB(192, 192, 192)

End Sub

Sub DrawCurrAlt(Pic As Object, val As Integer, Range As
Integer, base As Integer)
sh = Pic.ScaleHeight
```

```
sw = Pic.ScaleWidth

   Pic.Line (sw - 1, 0)-(sw - 1, sh), RGB(255, 255, 255)
   Pic.Line (sw - 1, sh - (sh * (val - base) / Range))-
(sw - 1, sh), RGB(128, 0, 0)
End Sub

Sub DrawCurrSpeed(Pic As Object, val As Integer)
Static maxspd As Integer

If val > maxspd Then maxspd = val

'VELOCITY = val

sh = Pic.ScaleHeight
sw = Pic.ScaleWidth

Pic.Line (0, 0)-(sw - 1, sh), RGB(255, 255, 255), BF
If (val > 55) Then
    Pic.Line (0, sh - (sh * val / 100))-(sw - 1, sh - (sh
* 55 / 100)), RGB(255, 0, 0), BF
    Pic.Line (0, sh)-(sw - 1, sh - (sh * 55 / 100)),
RGB(0, 128, 0), BF
Else
    Pic.Line (0, sh - (sh * val / 100))-(sw - 1, sh),
RGB(0, 128, 0), BF
End If

End Sub

Sub ScrollImg(Pic As Object)

      hDestDC = Pic.hDC
      hSrcDC = Pic.hDC
      nWidth = Pic.ScaleWidth
      nHeight = Pic.ScaleHeight

      picSrcX = 0
      picSrcY = 0

      picDestX = -1
      picDestY = 0

      ' Assign the SRCCOPY constant to the Raster
operation.
      dwRop = &HCC0020

      Suc = BitBlt(hDestDC, picDestX, picDestY, nWidth,
nHeight, hSrcDC, picSrcX, picSrcY, dwRop)
```

```
        Pic.Refresh

    End Sub

        Function StrToDeg(ByVal curstr As String) As Double

            Dim curdeg As Double
            Dim pe As Integer
            Dim deg As String, minz As String

            pe = InStr(curstr, ".")
            deg = Mid(curstr, 1, pe - 3)
            minz = Mid(curstr, pe - 2, Len(curstr) - (pe - 2))
            curdeg = deg + (minz / 60)

            StrToDeg = curdeg
        End function
```

Below is the code for the **imm4.bas** file, used when the KFMM variable is set to IMM4. The code is somewhat similar for the KF and MMAE options as well, so it is not being included in the appendix as well. Global variable initializations have been removed to simplify the code.

```
Attribute VB_Name = "IMM4"

Sub EKF_Setup()
   Dim Rows As Integer
   Dim Columns As Integer

   delta_k = 1

   '----- MATRIX I -----
   'Get size of array
   Call Mat.Mat_2D(i, Columns, Rows)
   'Load array with zeros
   Call Load_matrix_with_zeros(i, Rows, Columns)
   'Load corresponding cells with ones
   i(0, 0) = 1: i(1, 1) = 1: i(2, 2) = 1: i(3, 3) = 1:
i(4, 4) = 1: i(5, 5) = 1: i(6, 6) = 1: i(7, 7) = 1

   '----- MATRIX P -----
   'Get size of array
   Call Mat.Mat_2D(P1, Columns, Rows)
   'Load array with zeros
```

```
    Call Load_matrix_with_zeros(P1, Rows, Columns)
    'Load corresponding cells with ones
    CL_P1(0, 0) = 1: CL_P1(1, 1) = 1
    CL_P1(2, 2) = 1: CL_P1(3, 3) = 1
    CL_P1(4, 4) = 1: CL_P1(5, 5) = 1
    CL_P1(6, 6) = 1: CL_P1(7, 7) = 1

    CV_P1(0, 0) = 1: CV_P1(1, 1) = 1
    CV_P1(2, 2) = 1: CV_P1(3, 3) = 1
    CV_P1(4, 4) = 1: CV_P1(5, 5) = 1
    CV_P1(6, 6) = 1: CV_P1(7, 7) = 1

    CA_P1(0, 0) = 1: CA_P1(1, 1) = 1
    CA_P1(2, 2) = 1: CA_P1(3, 3) = 1
    CA_P1(4, 4) = 1: CA_P1(5, 5) = 1
    CA_P1(6, 6) = 1: CA_P1(7, 7) = 1

    CJ_P1(0, 0) = 1: CJ_P1(1, 1) = 1
    CJ_P1(2, 2) = 1: CJ_P1(3, 3) = 1
    CJ_P1(4, 4) = 1: CJ_P1(5, 5) = 1
    CJ_P1(6, 6) = 1: CJ_P1(7, 7) = 1

    CL_P = CL_P1
    CV_P = CV_P1
    CA_P = CA_P1
    CJ_P = CJ_P1

    CL_P2 = CL_P1
    CV_P2 = CV_P1
    CA_P2 = CA_P1
    CJ_P2 = CJ_P1

    '----- MATRIX A -----
    'Get size of array
    Call Mat.Mat_2D(CL_A, Columns, Rows)
    'Load array with zeros
    Call Load_matrix_with_zeros(CL_A, Rows, Columns)
    'Load corresponding cells with ones
    CL_A(0, 0) = 1: CL_A(1, 1) = 1: CL_A(2, 2) = 0:
CL_A(3, 3) = 0
    CL_A(4, 4) = 0: CL_A(5, 5) = 0: CL_A(6, 6) = 0:
CL_A(7, 7) = 0
    '----- MATRIX AT -----
    CL_AT = Mat.Transpose(CL_A)

    '----- MATRIX A -----
    'Get size of array
    Call Mat.Mat_2D(CV_A, Columns, Rows)
```

```
   'Load array with zeros
   Call Load_matrix_with_zeros(CV_A, Rows, Columns)
   'Load corresponding cells with ones
   CV_A(0, 0) = 1: CV_A(1, 1) = 1: CV_A(2, 2) = 1:
CV_A(3, 3) = 1
   CV_A(4, 4) = 0: CV_A(5, 5) = 0: CV_A(6, 6) = 0:
CV_A(7, 7) = 0
   CV_A(0, 2) = delta_k: CV_A(1, 3) = delta_k
   '----- MATRIX AT -----
   CV_AT = Mat.Transpose(CV_A)

   '----- MATRIX A -----
   'Get size of array
   Call Mat.Mat_2D(CA_A, Columns, Rows)
   'Load array with zeros
   Call Load_matrix_with_zeros(CA_A, Rows, Columns)
   'Load corresponding cells with ones
   CA_A(0, 0) = 1: CA_A(1, 1) = 1: CA_A(2, 2) = 1:
CA_A(3, 3) = 1
   CA_A(4, 4) = 1: CA_A(5, 5) = 1: CA_A(6, 6) = 0:
CA_A(7, 7) = 0
   CA_A(0, 2) = delta_k: CA_A(1, 3) = delta_k
   CA_A(2, 4) = delta_k: CA_A(3, 5) = delta_k
   CA_A(0, 4) = delta_k * delta_k: CA_A(1, 5) = delta_k *
delta_k
   '----- MATRIX AT -----
   CA_AT = Mat.Transpose(CA_A)

   '----- MATRIX A -----
   'Get size of array
   Call Mat.Mat_2D(CJ_A, Columns, Rows)
   'Load array with zeros
   Call Load_matrix_with_zeros(CJ_A, Rows, Columns)
   'Load corresponding cells with ones
   CJ_A(0, 0) = 1: CJ_A(1, 1) = 1: CJ_A(2, 2) = 1:
CJ_A(3, 3) = 1
   CJ_A(4, 4) = 1: CJ_A(5, 5) = 1: CJ_A(6, 6) = 1:
CJ_A(7, 7) = 1
   CJ_A(0, 2) = delta_k: CA_A(1, 3) = delta_k
   CJ_A(2, 4) = delta_k: CA_A(3, 5) = delta_k
   CJ_A(4, 6) = delta_k: CA_A(5, 7) = delta_k
   CJ_A(0, 4) = delta_k * delta_k: CA_A(1, 5) = delta_k *
delta_k
   CJ_A(2, 6) = delta_k * delta_k: CA_A(3, 7) = delta_k *
delta_k
   CJ_A(0, 6) = delta_k * delta_k * delta_k: CA_A(1, 7) =
delta_k * delta_k * delta_k
```

```
    '----- MATRIX AT -----
    CJ_AT = Mat.Transpose(CJ_A)

    Call Mat.Mat_2D(CL_Q, Columns, Rows)
'Get size of array
    Call Load_matrix_with_zeros(CL_Q, Rows, Columns)
'Load array with zeros
    Call Mat.Mat_2D(CV_Q, Columns, Rows)
'Get size of array
    Call Load_matrix_with_zeros(CV_Q, Rows, Columns)
'Load array with zeros
    Call Mat.Mat_2D(CA_Q, Columns, Rows)
'Get size of array
    Call Load_matrix_with_zeros(CA_Q, Rows, Columns)
'Load array with zeros
    Call Mat.Mat_2D(CJ_Q, Columns, Rows)
'Get size of array
    Call Load_matrix_with_zeros(CJ_Q, Rows, Columns)

CL_Q(0, 0) = 5.73333257747936E-12
CL_Q(0, 1) = 5.06472513631368E-13
CL_Q(1, 1) = 2.32803188514166E-12
CV_Q(0, 0) = 1.54505873523806E-14
CV_Q(0, 1) = -3.37231458507511E-16
CV_Q(1, 1) = 9.89653630420179E-15
CA_Q(0, 0) = 3.23023035267452E-15
CA_Q(0, 1) = -8.29674660571884E-17
CA_Q(1, 1) = 2.73268254605442E-15
CJ_Q(0, 0) = 5.50593224630771E-15
CJ_Q(0, 1) = -4.83965886722978E-17
CJ_Q(1, 1) = 5.70085282895857E-15
    'Common elementens in matrix
    CL_Q(1, 0) = CL_Q(0, 1)
    CV_Q(1, 0) = CV_Q(0, 1)
    CA_Q(1, 0) = CA_Q(0, 1)
    CJ_Q(1, 0) = CJ_Q(0, 1)

'Adjusting Q matrix exponents
For rrr = 0 To 1
    For ccc = 0 To 1
        CL_Q(rrr, ccc) = CL_Q(rrr, ccc) * 10 ^ 10
        CV_Q(rrr, ccc) = CV_Q(rrr, ccc) * 10 ^ 10
        CA_Q(rrr, ccc) = CA_Q(rrr, ccc) * 10 ^ 10
        CJ_Q(rrr, ccc) = CJ_Q(rrr, ccc) * 10 ^ 10
    Next
Next

   '----- MATRIX R ----- SAME FOR ALL KF MODELS
    Call Mat.Mat_2D(R1, Rows, Columns)
'Get size of array
```

```
   Call Load_matrix_with_zeros(R1, Rows, Columns)
  'Load array with zeros

  R1(0, 0) = 7.97784393584487E-15
  R1(0, 1) = -3.6150982125523E-16
  R1(0, 2) = -7.96313129788028E-15
  R1(0, 3) = 7.99189998051043E-16
  R1(0, 4) = -7.79397762003659E-15
  R1(0, 5) = 1.30862802760903E-15
  R1(0, 6) = -8.60219718382112E-15
  R1(0, 7) = 1.99726663401041E-15
  R1(1, 1) = 7.39844097826236E-15
  R1(1, 2) = 3.84542169608877E-16
  R1(1, 3) = -7.63900103442106E-15
  R1(1, 4) = 6.5744371963714E-16
  R1(1, 5) = -7.64805403347064E-15
  R1(1, 6) = 9.30109629011351E-16
  R1(1, 7) = -7.38230222896303E-15
  R1(2, 2) = 1.58719200618486E-14
  R1(2, 3) = -1.34289503218222E-15
  R1(2, 4) = 2.3749373334891E-14
  R1(2, 5) = -2.69242643566769E-15
  R1(2, 6) = 3.22124109145004E-14
  R1(2, 7) = -4.80850607902195E-15
  R1(3, 3) = 1.50332399488759E-14
  R1(3, 4) = -2.0607514624947E-15
  R1(3, 5) = 2.25453210231939E-14
  R1(3, 6) = -3.0269700164369E-15
  R1(3, 7) = 2.98363016022613E-14
  R1(4, 4) = 4.77904061115801E-14
  R1(4, 5) = -4.6680757308304E-15
  R1(4, 6) = 7.97050191442194E-14
  R1(4, 7) = -9.64753377256818E-15
  R1(5, 5) = 4.50610354660087E-14
  R1(5, 6) = -7.82692331295711E-15
  R1(5, 7) = 7.47891597973071E-14
  R1(6, 6) = 1.59676931638803E-13
  R1(6, 7) = -1.73480791759707E-14
  R1(7, 7) = 1.49616093724667E-13

  'Common elements in matrix
  R1(1, 0) = R1(0, 1)
  R1(2, 0) = R1(0, 2)
  R1(2, 1) = R1(1, 2)
  R1(3, 0) = R1(0, 3)
  R1(3, 1) = R1(1, 3)
  R1(3, 2) = R1(2, 3)
  R1(4, 0) = R1(0, 4)
```

```
R1(4, 1) = R1(1, 4)
R1(4, 2) = R1(2, 4)
R1(4, 3) = R1(3, 4)
R1(5, 0) = R1(0, 5)
R1(5, 1) = R1(1, 5)
R1(5, 2) = R1(2, 5)
R1(5, 3) = R1(3, 5)
R1(5, 4) = R1(4, 5)
R1(6, 0) = R1(0, 6)
R1(6, 1) = R1(1, 6)
R1(6, 2) = R1(2, 6)
R1(6, 3) = R1(3, 6)
R1(6, 4) = R1(4, 6)
R1(6, 5) = R1(5, 6)
R1(7, 0) = R1(0, 7)
R1(7, 1) = R1(1, 7)
R1(7, 2) = R1(2, 7)
R1(7, 3) = R1(3, 7)
R1(7, 4) = R1(4, 7)
R1(7, 5) = R1(5, 7)
R1(7, 6) = R1(6, 7)

'Adjusting R matrix exponents
For rrr = 0 To 7
   For ccc = 0 To 7
      R1(rrr, ccc) = R1(rrr, ccc) * 10 ^ 12
   Next
Next

   CL_R = R1
   CV_R = R1
   CA_R = R1
   CJ_R = R1

    'Get size of array
   Call Mat.Mat_2D(CL_W, Rows, Columns)
   'Load array with zeros
   Call Load_matrix_with_zeros(CL_W, Rows, Columns)
   'Load corresponding cells with ones
   CL_W(0, 0) = delta_k
   CL_W(1, 1) = delta_k
   '----- MATRIX AT -----
   CL_WT = Mat.Transpose(CL_W)

   'Get size of array
   Call Mat.Mat_2D(CV_W, Rows, Columns)
   'Load array with zeros
   Call Load_matrix_with_zeros(CV_W, Rows, Columns)
   'Load corresponding cells with ones
```

```
CV_W(0, 0) = delta_k * delta_k
CV_W(1, 1) = delta_k * delta_k
CV_W(2, 0) = delta_k
CV_W(3, 1) = delta_k
'----- MATRIX AT -----
CV_WT = Mat.Transpose(CV_W)

'Get size of array
Call Mat.Mat_2D(CA_W, Rows, Columns)
'Load array with zeros
Call Load_matrix_with_zeros(CA_W, Rows, Columns)
'Load corresponding cells with ones
CA_W(0, 0) = delta_k * delta_k * delta_k
CA_W(1, 1) = delta_k * delta_k * delta_k
CA_W(2, 0) = delta_k * delta_k
CA_W(3, 1) = delta_k * delta_k
CA_W(4, 0) = delta_k
CA_W(5, 1) = delta_k
'----- MATRIX AT -----
CA_WT = Mat.Transpose(CA_W)

'Get size of array
Call Mat.Mat_2D(CJ_W, Rows, Columns)
'Load array with zeros
Call Load_matrix_with_zeros(CJ_W, Rows, Columns)
'Load corresponding cells with ones
CJ_W(0, 0) = delta_k * delta_k * delta_k * delta_k
CJ_W(1, 1) = delta_k * delta_k * delta_k * delta_k
CJ_W(2, 0) = delta_k * delta_k * delta_k
CJ_W(3, 1) = delta_k * delta_k * delta_k
CJ_W(4, 0) = delta_k * delta_k
CJ_W(5, 1) = delta_k * delta_k
CJ_W(6, 0) = delta_k
CJ_W(7, 1) = delta_k
'----- MATRIX AT -----
CJ_WT = Mat.Transpose(CJ_W)

'----- MATRIX H -----
'Get size of array
Call Mat.Mat_2D(H, Rows, Columns)
'Load array with zeros
Call Load_matrix_with_zeros(H, Rows, Columns)
'Load corresponding cells with ones
H(0, 0) = 1: H(1, 1) = 1: H(2, 2) = 1: H(3, 3) = 1
H(4, 4) = 1: H(5, 5) = 1: H(6, 6) = 1: H(7, 7) = 1
'----- MATRIX AT -----
HT = Mat.Transpose(H)
```

```
    '----- MATRIX V -----
    'Get size of array
    Call Mat.Mat_2D(V, Rows, Columns)
    'Load array with zeros
    Call Load_matrix_with_zeros(V, Rows, Columns)
    'Load corresponding cells with ones
    V(0, 0) = 1: V(1, 1) = 1: V(2, 2) = 1: V(3, 3) = 1
    V(4, 4) = 1: V(5, 5) = 1: V(6, 6) = 1: V(7, 7) = 1
    '----- MATRIX VT -----
    VT = Mat.Transpose(V)

BT(0, 0) = 0.153846153846154
BT(0, 1) = 0.153846153846154
BT(0, 2) = 0.384615384615385
BT(0, 3) = 0.307692307692308
BT(1, 0) = 1.11265646731572E-02
BT(1, 1) = 0.47009735744089
BT(1, 2) = 0.304589707927677
BT(1, 3) = 0.214186369958275
BT(2, 0) = 1.44444444444444E-02
BT(2, 1) = 0.258888888888889
BT(2, 2) = 0.457777777777778
BT(2, 3) = 0.268888888888889
BT(3, 0) = 1.84162062615101E-03
BT(3, 1) = 0.243093922651934
BT(3, 2) = 0.50828729281768
BT(3, 3) = 0.246777163904236

    Set fs = CreateObject("Scripting.FileSystemObject")
     DATA_RESULTS = Mid(LOGNAME, 1, Len(LOGNAME) - 4)
     If SNAP2ROAD Then
         tmp = "withGIS"
     Else
         tmp = "noGIS"
     End If
    'Saving results file
    Set fout2 = fs.CreateTextFile(APP_PATH & "\data_
results_" & DATA_RESULTS & "_" & tmp & ".txt", True) '
create for writing

    Set fout3 = fs.OpenTextFile(APP_PATH & "\temp.txt", 2)
' open for writing

loop_cnt = 0
cnt = 0
filters = 4          'number of ekf filters

     CL_diff_avg(0) = 0
     CL_diff_avg(1) = 0
```

```
        CV_diff_avg(0) = 0
        CV_diff_avg(1) = 0
        CA_diff_avg(0) = 0
        CA_diff_avg(1) = 0
        CJ_diff_avg(0) = 0
        CJ_diff_avg(1) = 0
        IMM_diff_avg(0) = 0
        IMM_diff_avg(1) = 0
        Loop_count = 0

    End Sub

    Sub Update_Current_Location()

       Dim Rows As Integer
       Dim Columns As Integer
       Dim kfcl_x(3), kfcl_y(3), kfcv_x(3), kfcv_y(3),
    kfca_x(3), kfca_y(3), kfcj_x(3), kfcj_y(3), mmae_x(3),
    mmae_y(3)

        PERIOD = 5

        Loop_count = Loop_count + 1

    Locx_prev3 = Locx_prev2
    Locy_prev3 = Locy_prev2
    Locx_prev2 = Locx_prev1
    Locy_prev2 = Locy_prev1
    Locx_prev1 = Locx
    Locy_prev1 = Locy
    Locy = Kalman_Filters.Convert_Deg2Rad(curlat)
    Locx = Kalman_Filters.Convert_Deg2Rad(curlong)

    Vx_prev1 = Vx
    Vy_prev1 = Vy
    Vx = Locx - Locx_prev1
    Vy = Locy - Locy_prev1

    Ax_prev1 = Ax
    Ay_prev1 = Ay
    Ax = Vx - Vx_prev1
    Ay = Vy - Vy_prev1

    Jx_prev1 = Jx
    Jy_prev1 = Jy
    Jx = Ax - Ax_prev1
    Jy = Ay - Ay_prev1

    'Simple Estimations
    secv_x1 = secv_x1
```

```
    'Loading Z matrix with measured data
    Z(0, 0) = Locx: Z(1, 0) = Locy: Z(2, 0) = Vx: Z(3, 0)
= Vy
    Z(4, 0) = Ax:   Z(5, 0) = Ay:   Z(6, 0) = Jx: Z(7, 0)
= Jy

    'Initializing U1 Used in IMM calculations -->
U1=ui(k-1)  it gets re-calculated in step4 of the IMM
    U1(0) = 0.00001: U1(1) = 0.00001: U1(2) = 0.00001:
U1(3) = 0.00001

If (loop_cnt = PERIOD) Then

   CL_X_(0, 0) = Z(0, 0):   CL_X_(1, 0) = Z(1, 0):
CL_X_(2, 0) = Z(2, 0):     CL_X_(3, 0) = Z(3, 0)
   CL_X_(4, 0) = Z(4, 0):   CL_X_(5, 0) = Z(5, 0):
CL_X_(6, 0) = Z(6, 0):     CL_X_(7, 0) = Z(7, 0)
   CV_X_(0, 0) = Z(0, 0):   CV_X_(1, 0) = Z(1, 0):
CV_X_(2, 0) = Z(2, 0):     CV_X_(3, 0) = Z(3, 0)
   CV_X_(4, 0) = Z(4, 0):   CV_X_(5, 0) = Z(5, 0):
CV_X_(6, 0) = Z(6, 0):     CV_X_(7, 0) = Z(7, 0)
   CA_X_(0, 0) = Z(0, 0):   CA_X_(1, 0) = Z(1, 0):
CA_X_(2, 0) = Z(2, 0):     CA_X_(3, 0) = Z(3, 0)
   CA_X_(4, 0) = Z(4, 0):   CA_X_(5, 0) = Z(5, 0):
CA_X_(6, 0) = Z(6, 0):     CA_X_(7, 0) = Z(7, 0)
   CJ_X_(0, 0) = Z(0, 0):   CJ_X_(1, 0) = Z(1, 0):
CJ_X_(2, 0) = Z(2, 0):     CJ_X_(3, 0) = Z(3, 0)
   CJ_X_(4, 0) = Z(4, 0):   CJ_X_(5, 0) = Z(5, 0):
CJ_X_(6, 0) = Z(6, 0):     CJ_X_(7, 0) = Z(7, 0)

   estlong = curlong
   estlat = curlat

   ' Printing header of log file
    fout2.WriteLine ("units are in radians,
rad2feet=,20888089.218201744")
    fout2.WriteLine ( _
                  "Sections," & _
                  "GPS  0sec Xloc act," & _
                  "KFCL 1sec Xloc est," & "KFCV 1sec
Xloc est," & "KFCA 1sec Xloc est," & "KFCJ 1sec Xloc
est," & "MMAE 1sec Xloc est," & _
                  "KFCL 2sec Xloc est," & "KFCV 2sec
Xloc est," & "KFCA 2sec Xloc est," & "KFCJ 2sec Xloc
est," & "MMAE 2sec Xloc est," & _
                  "KFCL 3sec Xloc est," & "KFCV 3sec Xloc
est," & "KFCA 3sec Xloc est," & "KFCJ 3sec Xloc est," &
"MMAE 3sec Xloc est," & _
```

```
                    "------------------," & _
                    "GPS  0sec Yloc act," & _
                    "KFCL 1sec Yloc est," & "KFCV 1sec Yloc
est," & "KFCA 1sec Yloc est," & "KFCJ 1sec Yloc est," &
"MMAE 1sec Yloc est," & _
                    "KFCL 2sec Yloc est," & "KFCV 2sec Yloc
est," & "KFCA 2sec Yloc est," & "KFCJ 2sec Yloc est," &
"MMAE 2sec Yloc est," & _
                    "KFCL 3sec Yloc est," & "KFCV 3sec Yloc
est," & "KFCA 3sec Yloc est," & "KFCJ 3sec Yloc est," &
"MMAE 3sec Yloc est," & _
                    model_text & _
                     "")

   'Predict next position with updated filter
   Call CL_model_predict
   Call CV_model_predict
   Call CA_model_predict
   Call CJ_model_predict

ElseIf (loop_cnt > PERIOD) Then

    If (loop_cnt = PERIOD * 5) Then
        xloc = Locx
        yloc = Locy
    End If

B_cnt = 1

If SHOWLINE Then

        'Display Previous Location
        Set objPin = objMap.FindPushpin("Previous
Location")
        objPin.Delete
        Set objLoc = objMap.GetLocation(Kalman_Filters.
Convert_Rad2Deg(Locy_prev1), Kalman_Filters.
Convert_Rad2Deg(Locx_prev1))
        objMap.AddPushpin objLoc, "Previous Location"

End If

        'Display Current Location
        Set objPin = objMap.FindPushpin("Current
Location")
        objPin.Delete
        Set objLoc = objMap.GetLocation(curlat, curlong)
        objMap.AddPushpin objLoc, "Current Location"
```

```
        Set objCurLoc = objMap.FindPushpin("Current
Location")
        objCurLoc.Symbol = 30 '82
        objCurLoc.Location.GoTo

If CALCOFFSET Then
    Call Calc_offset.Start(curlong, curlat, Kalman_
Filters.Convert_Rad2Deg(Locx_prev1), Kalman_Filters.
Convert_Rad2Deg(Locy_prev1))
Else
    offset_lat = 0
    offset_lon = 0
End If

    '----- EKF Correct Step -----
   Call CL_model_correct   'Correct previous prediction
with obtained data
   Call CV_model_correct   'Correct previous prediction
with obtained data
   Call CA_model_correct   'Correct previous prediction
with obtained data
   Call CJ_model_correct   'Correct previous prediction
with obtained data

    '----- IMM Steps 1,2,3 -----
If (IMM_STEPS) Then
    Call MM_filter_part1
End If

    '#### Saving variable's values before next predict
step ######################################

    If (loop_cnt > PERIOD * 5) Then
        '----------------------------------------
        If (SYSTEMLOOP) Then   'Change to TRUE to go into
this extended loop of MM

    CL_h_loop_(0, 0) = CL_h_(0, 0): CL_h_loop_(1, 0) =
CL_h_(1, 0): CL_h_loop_(2, 0) = CL_h_(2, 0): CL_h_loop_
(3, 0) = CL_h_(3, 0)
    CL_h_loop_(4, 0) = CL_h_(4, 0): CL_h_loop_(5, 0) =
CL_h_(5, 0): CL_h_loop_(6, 0) = CL_h_(6, 0): CL_h_loop_
(7, 0) = CL_h_(7, 0)
    CV_h_loop_(0, 0) = CV_h_(0, 0): CV_h_loop_(1, 0) =
CV_h_(1, 0): CV_h_loop_(2, 0) = CV_h_(2, 0): CV_h_loop_
(3, 0) = CV_h_(3, 0)
    CV_h_loop_(4, 0) = CV_h_(4, 0): CV_h_loop_(5, 0) =
CV_h_(5, 0): CV_h_loop_(6, 0) = CV_h_(6, 0): CV_h_loop_
(7, 0) = CV_h_(7, 0)
```

```
    CA_h_loop_(0, 0) = CA_h_(0, 0): CA_h_loop_(1, 0) =
CA_h_(1, 0): CA_h_loop_(2, 0) = CA_h_(2, 0): CA_h_loop_
(3, 0) = CA_h_(3, 0)
    CA_h_loop_(4, 0) = CA_h_(4, 0): CA_h_loop_(5, 0) =
CA_h_(5, 0): CA_h_loop_(6, 0) = CA_h_(6, 0): CA_h_loop_
(7, 0) = CA_h_(7, 0)
    CJ_h_loop_(0, 0) = CJ_h_(0, 0): CJ_h_loop_(1, 0) =
CJ_h_(1, 0): CJ_h_loop_(2, 0) = CJ_h_(2, 0): CJ_h_loop_
(3, 0) = CJ_h_(3, 0)
    CJ_h_loop_(4, 0) = CJ_h_(4, 0): CJ_h_loop_(5, 0) =
CJ_h_(5, 0): CJ_h_loop_(6, 0) = CJ_h_(6, 0): CJ_h_loop_
(7, 0) = CJ_h_(7, 0)

    CL_X_loop_(0, 0) = CL_X_(0, 0): CL_X_loop_(1, 0) =
CL_X_(1, 0): CL_X_loop_(2, 0) = CL_X_(2, 0): CL_X_loop_
(3, 0) = CL_X_(3, 0)
    CL_X_loop_(4, 0) = CL_X_(4, 0): CL_X_loop_(5, 0) =
CL_X_(5, 0): CL_X_loop_(6, 0) = CL_X_(6, 0): CL_X_loop_
(7, 0) = CL_X_(7, 0)
    CV_X_loop_(0, 0) = CV_X_(0, 0): CV_X_loop_(1, 0) =
CV_X_(1, 0): CV_X_loop_(2, 0) = CV_X_(2, 0): CV_X_loop_
(3, 0) = CV_X_(3, 0)
    CV_X_loop_(4, 0) = CV_X_(4, 0): CV_X_loop_(5, 0) =
CV_X_(5, 0): CV_X_loop_(6, 0) = CV_X_(6, 0): CV_X_loop_
(7, 0) = CV_X_(7, 0)
    CA_X_loop_(0, 0) = CA_X_(0, 0): CA_X_loop_(1, 0) =
CA_X_(1, 0): CA_X_loop_(2, 0) = CA_X_(2, 0): CA_X_loop_
(3, 0) = CA_X_(3, 0)
    CA_X_loop_(4, 0) = CA_X_(4, 0): CA_X_loop_(5, 0) =
CA_X_(5, 0): CA_X_loop_(6, 0) = CA_X_(6, 0): CA_X_loop_
(7, 0) = CA_X_(7, 0)
    CJ_X_loop_(0, 0) = CJ_X_(0, 0): CJ_X_loop_(1, 0) =
CJ_X_(1, 0): CJ_X_loop_(2, 0) = CJ_X_(2, 0): CJ_X_loop_
(3, 0) = CJ_X_(3, 0)
    CJ_X_loop_(4, 0) = CJ_X_(4, 0): CJ_X_loop_(5, 0) =
CJ_X_(5, 0): CJ_X_loop_(6, 0) = CJ_X_(6, 0): CJ_X_loop_
(7, 0) = CJ_X_(7, 0)

    CL_P_loop = CL_P
    CV_P_loop = CV_P
    CA_P_loop = CA_P
    CJ_P_loop = CJ_P

    CL_HP2HT_loop = CL_HP2HT
    CV_HP2HT_loop = CV_HP2HT
    CA_HP2HT_loop = CA_HP2HT
    CJ_HP2HT_loop = CJ_HP2HT
```

```
    Z_loop(0, 0) = Z(0, 0): Z_loop(1, 0) = Z(1, 0): Z_
loop(2, 0) = Z(2, 0): Z_loop(3, 0) = Z(3, 0)
    Z_loop(4, 0) = Z(4, 0): Z_loop(5, 0) = Z(5, 0): Z_
loop(6, 0) = Z(6, 0): Z_loop(7, 0) = Z(7, 0)

    cb_loop(0) = cb(0): cb_loop(1) = cb(1): cb_loop(2) =
cb(2): cb_loop(3) = cb(3)

        End If
        '----------------------------------------
    End If

   '----- EKF Predict Step -----
   Call CL_model_predict   'Predict next position with
updated filter
   Call CV_model_predict   'Predict next position with
updated filter
   Call CA_model_predict   'Predict next position with
updated filter
   Call CJ_model_predict   'Predict next position with
updated filter

   '----- IMM Steps 4,5 -----
If (IMM_STEPS) Then
    Call MM_filter_part2
End If

            kfcl_x(1) = Abs(CL_h_(0, 0))
            kfcv_x(1) = Abs(CV_h_(0, 0))
            kfca_x(1) = Abs(CA_h_(0, 0))
            kfcj_x(1) = Abs(CJ_h_(0, 0))
            mmae_x(1) = Abs(XM(0))

            kfcl_y(1) = Abs(CL_h_(1, 0))
            kfcv_y(1) = Abs(CV_h_(1, 0))
            kfca_y(1) = Abs(CA_h_(1, 0))
            kfcj_y(1) = Abs(CJ_h_(1, 0))
            mmae_y(1) = Abs(XM(1))

    If (loop_cnt > PERIOD * 5) Then
        '----------------------------------------
        If (SYSTEMLOOP) Then   'Change to TRUE to go into
this extended loop of MM
    delta_k_loop = 3

   '----- EKF Predict Step -----
   Call CL_model_predict_loop   'Predict next position
with updated filter
```

```
   Call CV_model_predict_loop   'Predict next position
with updated filter
   Call CA_model_predict_loop   'Predict next position
with updated filter
   Call CJ_model_predict_loop   'Predict next position
with updated filter

   '----- IMM Steps 4,5 -----
    If (IMM_STEPS) Then
        Call MM_filter_part2_loop
    End If
    delta_k_loop = 1

            B_cnt = 3 '3sec ahead
            kfcl_x(B_cnt) = Abs(CL_h_loop_(0, 0))
            kfcv_x(B_cnt) = Abs(CV_h_loop_(0, 0))
            kfca_x(B_cnt) = Abs(CA_h_loop_(0, 0))
            kfcj_x(B_cnt) = Abs(CJ_h_loop_(0, 0))
            mmae_x(B_cnt) = Abs(XM_loop(0))

            kfcl_y(B_cnt) = Abs(CL_h_loop_(1, 0))
            kfcv_y(B_cnt) = Abs(CV_h_loop_(1, 0))
            kfca_y(B_cnt) = Abs(CA_h_loop_(1, 0))
            kfcj_y(B_cnt) = Abs(CJ_h_loop_(1, 0))
            mmae_y(B_cnt) = Abs(XM_loop(1))

        End If
    End If

'keeping track of previous values
CL_Locx_prev(3) = CL_Locx_prev(2)
CL_Locy_prev(3) = CL_Locy_prev(2)
CL_Locx_prev(2) = CL_Locx_prev(1)
CL_Locy_prev(2) = CL_Locy_prev(1)
CL_Locx_prev(1) = CL_Locx
CL_Locy_prev(1) = CL_Locy
CL_Locx = kfcl_x(sec_ahead)
CL_Locy = kfcl_y(sec_ahead)

CV_Locx_prev(3) = CV_Locx_prev(2)
CV_Locy_prev(3) = CV_Locy_prev(2)
CV_Locx_prev(2) = CV_Locx_prev(1)
CV_Locy_prev(2) = CV_Locy_prev(1)
CV_Locx_prev(1) = CV_Locx
CV_Locy_prev(1) = CV_Locy
CV_Locx = kfcv_x(sec_ahead)
CV_Locy = kfcv_y(sec_ahead)
```

```
CA_Locx_prev(3) = CA_Locx_prev(2)
CA_Locy_prev(3) = CA_Locy_prev(2)
CA_Locx_prev(2) = CA_Locx_prev(1)
CA_Locy_prev(2) = CA_Locy_prev(1)
CA_Locx_prev(1) = CA_Locx
CA_Locy_prev(1) = CA_Locy
CA_Locx = kfca_x(sec_ahead)
CA_Locy = kfca_y(sec_ahead)

CJ_Locx_prev(3) = CJ_Locx_prev(2)
CJ_Locy_prev(3) = CJ_Locy_prev(2)
CJ_Locx_prev(2) = CJ_Locx_prev(1)
CJ_Locy_prev(2) = CJ_Locy_prev(1)
CJ_Locx_prev(1) = CJ_Locx
CJ_Locy_prev(1) = CJ_Locy
CJ_Locx = kfcj_x(sec_ahead)
CJ_Locy = kfcj_y(sec_ahead)

IMM_Locx_prev(3) = IMM_Locx_prev(2)
IMM_Locy_prev(3) = IMM_Locy_prev(2)
IMM_Locx_prev(2) = IMM_Locx_prev(1)
IMM_Locy_prev(2) = IMM_Locy_prev(1)
IMM_Locx_prev(1) = IMM_Locx
IMM_Locy_prev(1) = IMM_Locy
IMM_Locx = mmae_x(sec_ahead)
IMM_Locy = mmae_y(sec_ahead)

If (Loop_count >= REC_NUM And Loop_count <= STOP_NUM)
Then
If (Not CL_Locx_prev(sec_ahead) = 0) Then

    rad2meters = 20888089.2182017 * 0.3048
    CL_diff = ((Abs(Abs(Locx) - Abs(CL_Locx_prev(sec_
ahead))) + Abs(Abs(Locy) - Abs(CL_Locy_prev(sec_ahead))))
/ 2) * rad2meters
    CV_diff = ((Abs(Abs(Locx) - Abs(CV_Locx_prev(sec_
ahead))) + Abs(Abs(Locy) - Abs(CV_Locy_prev(sec_ahead))))
/ 2) * rad2meters
    CA_diff = ((Abs(Abs(Locx) - Abs(CA_Locx_prev(sec_
ahead))) + Abs(Abs(Locy) - Abs(CA_Locy_prev(sec_ahead))))
/ 2) * rad2meters
    CJ_diff = ((Abs(Abs(Locx) - Abs(CJ_Locx_prev(sec_
ahead))) + Abs(Abs(Locy) - Abs(CJ_Locy_prev(sec_ahead))))
/ 2) * rad2meters
    IMM_diff = ((Abs(Abs(Locx) - Abs(IMM_Locx_prev(sec_
ahead))) + Abs(Abs(Locy) - Abs(IMM_Locy_prev(sec_
ahead)))) / 2) * rad2meters
```

```
    CL_diff_avg(0) = CL_diff_avg(0) + CL_diff
    CL_diff_avg(1) = CL_diff_avg(1) + 1
    CL_diff2 = CL_diff_avg(0) / CL_diff_avg(1)
    CV_diff_avg(0) = CV_diff_avg(0) + CV_diff
    CV_diff_avg(1) = CV_diff_avg(1) + 1
    CV_diff2 = CV_diff_avg(0) / CV_diff_avg(1)
    CA_diff_avg(0) = CA_diff_avg(0) + CA_diff
    CA_diff_avg(1) = CA_diff_avg(1) + 1
    CA_diff2 = CA_diff_avg(0) / CA_diff_avg(1)
    CJ_diff_avg(0) = CJ_diff_avg(0) + CJ_diff
    CJ_diff_avg(1) = CJ_diff_avg(1) + 1
    CJ_diff2 = CJ_diff_avg(0) / CJ_diff_avg(1)
    IMM_diff_avg(0) = IMM_diff_avg(0) + IMM_diff
    IMM_diff_avg(1) = IMM_diff_avg(1) + 1
    IMM_diff2 = IMM_diff_avg(0) / IMM_diff_avg(1)

    Form1.Label3.Caption = Loop_count & "   CL=" & CL_
diff2 & "   CV=" & CV_diff2 & "   CA=" & CA_diff2 & "
CJ=" & CJ_diff2 & "   IMM=" & IMM_diff2

    IMM_diffx = Abs(Abs(CL_Locx_prev(sec_ahead)) -
Abs(xloc)) * rad2meters
    IMM_diffy = Abs(Abs(CL_Locy_prev(sec_ahead)) -
Abs(yloc)) * rad2meters

    GPSx = Abs(Abs(Locx) - Abs(xloc)) * rad2meters
    GPSy = Abs(Abs(Locy) - Abs(yloc)) * rad2meters

    fout3.WriteLine (Loop_count & "," & CL_diff & "," &
CV_diff & "," & CA_diff & "," & CJ_diff & "," & IMM_diff
& ",,," & IMM_diffx & "," & IMM_diffy & "," & GPSx & ","
& GPSy)

End If
End If

    model_text = CL_diff & "," & CV_diff & "," & CA_diff
& "," & CJ_diff & "," & IMM_diff & "," & _
                CL_diff2 & "," & CV_diff2 & "," & CA_diff2
& "," & CJ_diff2 & "," & IMM_diff2 & ","

        fout2.WriteLine ( _
                         section_no & "," & _
                         Abs(Z(0, 0)) & "," & _
                         kfcl_x(1) & "," & kfcv_x(1) &
"," & kfca_x(1) & "," & kfcj_x(1) & "," & _
                         mmae_x(1) & "," & _
```

```
                              kfcl_x(2) & "," & kfcv_x(2) &
"," & kfca_x(2) & "," & kfcj_x(2) & "," & _
                              mmae_x(2) & "," & _
                              kfcl_x(3) & "," & kfcv_x(3) &
"," & kfca_x(3) & "," & kfcj_x(3) & "," & _
                              mmae_x(3) & "," & _
                              "," & _
                              Abs(Z(1, 0)) & "," & _
                              kfcl_y(1) & "," & kfcv_y(1) &
"," & kfca_y(1) & "," & kfcj_y(1) & "," & _
                              mmae_y(1) & "," & _
                              kfcl_y(2) & "," & kfcv_y(2) &
"," & kfca_y(2) & "," & kfcj_y(2) & "," & _
                              mmae_y(2) & "," & _
                              kfcl_y(3) & "," & kfcv_y(3) &
"," & kfca_y(3) & "," & kfcj_y(3) & "," & _
                              mmae_y(3) & "," & _
                              "," & model_text & _
                              "")

End If

   loop_cnt = loop_cnt + 1

End Sub

Sub CL_model_correct()
   CL_HP2HT = Mat.Multiply(Mat.Multiply(H, CL_P2), HT) ':
aaa = display_matrix(CL_HP2HT, "CL_HP2HT")
   CL_VRVT = Mat.Multiply(Mat.Multiply(V, CL_R), VT) ':
aaa = display_matrix(CL_VRVT, "CL_VRVT")
   CL_S = Mat.Add(CL_HP2HT, CL_VRVT) ': aaa = display_
matrix(CL_S, "CL_S")
   CL_Sinv = Mat.Inv(CL_S)
   CL_P2HT = Mat.Multiply(CL_P2, HT)
   CL_k = Mat.Multiply(CL_P2HT, CL_Sinv)
   CL_k = Dampen_K(CL_k, "CL")
   CL_X_tmp = Mat.Add(CL_h_, Mat.Multiply(CL_k, Mat.
Subtract(Z, CL_h_)))
   CL_P = Mat.Multiply(Mat.Subtract(i, Mat.Multiply(CL_k,
H)), CL_P2)
   CL_X(0, 0) = CL_X_tmp(0, 0):    CL_X(1, 0) = CL_X_
tmp(1, 0):    CL_X(2, 0) = CL_X_tmp(2, 0):    CL_X(3, 0)
= CL_X_tmp(3, 0)
   CL_X(4, 0) = CL_X_tmp(4, 0):    CL_X(5, 0) = CL_X_
tmp(5, 0):    CL_X(6, 0) = CL_X_tmp(6, 0):    CL_X(7, 0)
= CL_X_tmp(7, 0)
```

```
   CL_X_(0, 0) = CL_X(0, 0):          CL_X_(1, 0) = CL_X(1,
0):        CL_X_(2, 0) = CL_X(2, 0):        CL_X_(3, 0) =
CL_X(3, 0)
   CL_X_(4, 0) = CL_X(4, 0):          CL_X_(5, 0) = CL_X(5,
0):        CL_X_(6, 0) = CL_X(6, 0):        CL_X_(7, 0) =
CL_X(7, 0)

End Sub

Sub CL_model_predict()

   CL_X(0, 0) = CL_X_(0, 0)
   CL_X(1, 0) = CL_X_(1, 0)
   CL_X(2, 0) = 0 'CL_X_(2, 0)
   CL_X(3, 0) = 0 'CL_X_(3, 0)
   CL_X(4, 0) = 0 'CL_X_(4, 0)
   CL_X(5, 0) = 0 'CL_X_(5, 0)
   CL_X(6, 0) = 0 'CL_X_(6, 0)
   CL_X(7, 0) = 0 'CL_X_(7, 0)

   CL_APAT = Mat.Multiply(Mat.Multiply(CL_A, CL_P), CL_AT)
   CL_WQWT = Mat.Multiply(Mat.Multiply(CL_W, CL_Q), CL_WT)
   CL_P2 = Mat.Add(CL_APAT, CL_WQWT)

   CL_h_(0, 0) = CL_X(0, 0): CL_h_(1, 0) = CL_X(1, 0):
CL_h_(2, 0) = CL_X(2, 0): CL_h_(3, 0) = CL_X(3, 0)
   CL_h_(4, 0) = CL_X(4, 0): CL_h_(5, 0) = CL_X(5, 0):
CL_h_(6, 0) = CL_X(6, 0): CL_h_(7, 0) = CL_X(7, 0)

End Sub

Sub CV_model_correct()

   CV_HP2HT = Mat.Multiply(Mat.Multiply(H, CV_P2), HT)
   CV_VRVT = Mat.Multiply(Mat.Multiply(V, CV_R), VT)
   CV_S = Mat.Add(CV_HP2HT, CV_VRVT)
   CV_Sinv = Mat.Inv(CV_S)
   CV_P2HT = Mat.Multiply(CV_P2, HT)
   CV_k = Mat.Multiply(CV_P2HT, CV_Sinv)
   CV_k = Dampen_K(CV_k, "CV")
   CV_X_tmp = Mat.Add(CV_h_, Mat.Multiply(CV_k, Mat.
Subtract(Z, CV_h_)))
   CV_P = Mat.Multiply(Mat.Subtract(i, Mat.Multiply(CV_k,
H)), CV_P2)

   CV_X(0, 0) = CV_X_tmp(0, 0):     CV_X(1, 0) = CV_X_
tmp(1, 0):    CV_X(2, 0) = CV_X_tmp(2, 0):    CV_X(3, 0)
= CV_X_tmp(3, 0)
```

```
   CV_X(4, 0) = CV_X_tmp(4, 0):    CV_X(5, 0) = CV_X_
tmp(5, 0):    CV_X(6, 0) = CV_X_tmp(6, 0):    CV_X(7, 0)
= CV_X_tmp(7, 0)
   CV_X_(0, 0) = CV_X(0, 0):       CV_X_(1, 0) = CV_X(1,
0):       CV_X_(2, 0) = CV_X(2, 0):       CV_X_(3, 0) =
CV_X(3, 0)
   CV_X_(4, 0) = CV_X(4, 0):       CV_X_(5, 0) = CV_X(5,
0):       CV_X_(6, 0) = CV_X(6, 0):       CV_X_(7, 0) =
CV_X(7, 0)

End Sub

Sub CV_model_predict()

   CV_X(0, 0) = CV_X_(0, 0) + CV_X_(2, 0) * delta_k
   CV_X(1, 0) = CV_X_(1, 0) + CV_X_(3, 0) * delta_k
   CV_X(2, 0) = CV_X_(2, 0)
   CV_X(3, 0) = CV_X_(3, 0)
   CV_X(4, 0) = 0 'CV_X_(4, 0)
   CV_X(5, 0) = 0 'CV_X_(5, 0)
   CV_X(6, 0) = 0 'CV_X_(6, 0)
   CV_X(7, 0) = 0 'CV_X_(7, 0)

   CV_APAT = Mat.Multiply(Mat.Multiply(CV_A, CV_P),
CV_AT)
   CV_WQWT = Mat.Multiply(Mat.Multiply(CV_W, CV_Q),
CV_WT)
   CV_P2 = Mat.Add(CV_APAT, CV_WQWT)

   CV_h_(0, 0) = CV_X(0, 0): CV_h_(1, 0) = CV_X(1, 0):
CV_h_(2, 0) = CV_X(2, 0): CV_h_(3, 0) = CV_X(3, 0)
   CV_h_(4, 0) = CV_X(4, 0): CV_h_(5, 0) = CV_X(5, 0):
CV_h_(6, 0) = CV_X(6, 0): CV_h_(7, 0) = CV_X(7, 0)

End Sub

Sub CA_model_correct()

   CA_HP2HT = Mat.Multiply(Mat.Multiply(H, CA_P2), HT)
   CA_VRVT = Mat.Multiply(Mat.Multiply(V, CA_R), VT)
   CA_S = Mat.Add(CA_HP2HT, CA_VRVT)
   CA_Sinv = Mat.Inv(CA_S)
   CA_P2HT = Mat.Multiply(CA_P2, HT)
   CA_k = Mat.Multiply(CA_P2HT, CA_Sinv)
   CA_k = Dampen_K(CA_k, "CA")

   CA_X_tmp = Mat.Add(CA_h_, Mat.Multiply(CA_k, Mat.
Subtract(Z, CA_h_)))
```

```
   CA_P = Mat.Multiply(Mat.Subtract(i, Mat.Multiply(CA_k,
H)), CA_P2)

   CA_X(0, 0) = CA_X_tmp(0, 0):     CA_X(1, 0) = CA_X_
tmp(1, 0):     CA_X(2, 0) = CA_X_tmp(2, 0):     CA_X(3, 0)
= CA_X_tmp(3, 0)
   CA_X(4, 0) = CA_X_tmp(4, 0):     CA_X(5, 0) = CA_X_
tmp(5, 0):     CA_X(6, 0) = CA_X_tmp(6, 0):     CA_X(7, 0)
= CA_X_tmp(7, 0)
   CA_X_(0, 0) = CA_X(0, 0):        CA_X_(1, 0) = CA_X(1,
0):      CA_X_(2, 0) = CA_X(2, 0):        CA_X_(3, 0) =
CA_X(3, 0)
   CA_X_(4, 0) = CA_X(4, 0):        CA_X_(5, 0) = CA_X(5,
0):      CA_X_(6, 0) = CA_X(6, 0):        CA_X_(7, 0) =
CA_X(7, 0)

End Sub

Sub CA_model_predict()

   CA_X(0, 0) = CA_X_(0, 0) + CA_X_(2, 0) * delta_k + (1
/ 2) * CA_X_(4, 0) * delta_k * delta_k
   CA_X(1, 0) = CA_X_(1, 0) + CA_X_(3, 0) * delta_k + (1
/ 2) * CA_X_(5, 0) * delta_k * delta_k
   CA_X(2, 0) = CA_X_(2, 0) + CA_X_(4, 0) * delta_k
   CA_X(3, 0) = CA_X_(3, 0) + CA_X_(5, 0) * delta_k
   CA_X(4, 0) = CA_X_(4, 0)
   CA_X(5, 0) = CA_X_(5, 0)
   CA_X(6, 0) = 0 'CA_X_(6, 0)
   CA_X(7, 0) = 0 'CA_X_(7, 0)

   CA_APAT = Mat.Multiply(Mat.Multiply(CA_A, CA_P),
CA_AT)
   CA_WQWT = Mat.Multiply(Mat.Multiply(CA_W, CA_Q),
CA_WT)
   CA_P2 = Mat.Add(CA_APAT, CA_WQWT)

   CA_h_(0, 0) = CA_X(0, 0): CA_h_(1, 0) = CA_X(1, 0):
CA_h_(2, 0) = CA_X(2, 0): CA_h_(3, 0) = CA_X(3, 0)
   CA_h_(4, 0) = CA_X(4, 0): CA_h_(5, 0) = CA_X(5, 0):
CA_h_(6, 0) = CA_X(6, 0): CA_h_(7, 0) = CA_X(7, 0)

End Sub

Sub CJ_model_correct()

   CJ_HP2HT = Mat.Multiply(Mat.Multiply(H, CJ_P2), HT)
   CJ_VRVT = Mat.Multiply(Mat.Multiply(V, CJ_R), VT)
```

```
    CJ_S = Mat.Add(CJ_HP2HT, CJ_VRVT)
    CJ_Sinv = Mat.Inv(CJ_S)
    CJ_P2HT = Mat.Multiply(CJ_P2, HT)
    CJ_k = Mat.Multiply(CJ_P2HT, CJ_Sinv)
    CJ_k = Dampen_K(CJ_k, "CJ")

    CJ_X_tmp = Mat.Add(CJ_h_, Mat.Multiply(CJ_k, Mat.
Subtract(Z, CJ_h_)))
    CJ_P = Mat.Multiply(Mat.Subtract(i, Mat.Multiply(CJ_k,
H)), CJ_P2)
    CJ_X(0, 0) = CJ_X_tmp(0, 0):     CJ_X(1, 0) = CJ_X_
tmp(1, 0):     CJ_X(2, 0) = CJ_X_tmp(2, 0):     CJ_X(3, 0)
= CJ_X_tmp(3, 0)
    CJ_X(4, 0) = CJ_X_tmp(4, 0):     CJ_X(5, 0) = CJ_X_
tmp(5, 0):     CJ_X(6, 0) = CJ_X_tmp(6, 0):     CJ_X(7, 0)
= CJ_X_tmp(7, 0)
    CJ_X_(0, 0) = CJ_X(0, 0):        CJ_X_(1, 0) = CJ_X(1,
0):       CJ_X_(2, 0) = CJ_X(2, 0):        CJ_X_(3, 0) =
CJ_X(3, 0)
    CJ_X_(4, 0) = CJ_X(4, 0):        CJ_X_(5, 0) = CJ_X(5,
0):       CJ_X_(6, 0) = CJ_X(6, 0):        CJ_X_(7, 0) =
CJ_X(7, 0)

End Sub

Sub CJ_model_predict()

    CJ_X(0, 0) = CJ_X_(0, 0) + CJ_X_(2, 0) * delta_k + (1
/ 2) * CJ_X_(4, 0) * delta_k * delta_k + (1 / 6) * CJ_X_
(6, 0) * delta_k * delta_k * delta_k
    CJ_X(1, 0) = CJ_X_(1, 0) + CJ_X_(3, 0) * delta_k + (1
/ 2) * CJ_X_(5, 0) * delta_k * delta_k + (1 / 6) * CJ_X_
(7, 0) * delta_k * delta_k * delta_k
    CJ_X(2, 0) = CJ_X_(2, 0) + CJ_X_(4, 0) * delta_k + (1
/ 2) * CJ_X_(6, 0) * delta_k * delta_k
    CJ_X(3, 0) = CJ_X_(3, 0) + CJ_X_(5, 0) * delta_k + (1
/ 2) * CJ_X_(7, 0) * delta_k * delta_k
    CJ_X(4, 0) = CJ_X_(4, 0) + CJ_X_(6, 0) * delta_k
    CJ_X(5, 0) = CJ_X_(5, 0) + CJ_X_(7, 0) * delta_k
    CJ_X(6, 0) = CJ_X_(6, 0)
    CJ_X(7, 0) = CJ_X_(7, 0)

    CJ_APAT = Mat.Multiply(Mat.Multiply(CJ_A, CJ_P),
CJ_AT)
    CJ_WQWT = Mat.Multiply(Mat.Multiply(CJ_W, CJ_Q),
CJ_WT)
    CJ_P2 = Mat.Add(CJ_APAT, CJ_WQWT)
```

```
   CJ_h_(0, 0) = CJ_X(0, 0): CJ_h_(1, 0) = CJ_X(1, 0):
CJ_h_(2, 0) = CJ_X(2, 0): CJ_h_(3, 0) = CJ_X(3, 0)
   CJ_h_(4, 0) = CJ_X(4, 0): CJ_h_(5, 0) = CJ_X(5, 0):
CJ_h_(6, 0) = CJ_X(6, 0): CJ_h_(7, 0) = CJ_X(7, 0)

End Sub

Sub MM_filter_part1()
    Dim U(3, 3) As Double
    Dim CL_X0(7, 0) As Double, CV_X0(7, 0) As Double,
CA_X0(7, 0) As Double, CJ_X0(7, 0) As Double
    Dim CL_errj0_(7, 0) As Double, CV_errj0_(7, 0) As
Double, CA_errj0_(7, 0) As Double, CJ_errj0_(7, 0) As
Double
    Dim CL_errj0() As Double, CV_errj0() As Double, CA_
errj0() As Double, CJ_errj0() As Double
    Dim CL_P0(7, 7) As Double, CV_P0(7, 7) As Double,
CA_P0(7, 7) As Double, CJ_P0(7, 7) As Double

    '--- IMM Step 1 --- Calculation of the mixing
probabilities
    For col = 0 To 3

        cb(col) = 0
        For row = 0 To 3
            cb(col) = cb(col) + BT(row, col) * U1(row)
        Next row

        For row = 0 To 3
            U(row, col) = (1 / cb(col)) * BT(row, col) *
            U1(row)

    'Preventing it going to zero
            If (U(row, col) <= 0) Then
                U(row, col) = 0.0000000001
            End If

        Next row

    Next col

    '--- IMM Step 2 --- Mixing
    For r = 0 To 7
        CL_X0(r, 0) = CL_X_(r, 0) * U(0, 0) + CV_X_(r, 0)
* U(1, 0) + CA_X_(r, 0) * U(2, 0) + CJ_X_(r, 0) * U(3, 0)
        CV_X0(r, 0) = CL_X_(r, 0) * U(0, 1) + CV_X_(r, 0)
* U(1, 1) + CA_X_(r, 0) * U(2, 1) + CJ_X_(r, 0) * U(3, 1)
        CA_X0(r, 0) = CL_X_(r, 0) * U(0, 2) + CV_X_(r, 0)
* U(1, 2) + CA_X_(r, 0) * U(2, 2) + CJ_X_(r, 0) * U(3, 2)
```

```
        CJ_X0(r, 0) = CL_X_(r, 0) * U(0, 3) + CV_X_(r, 0)
* U(1, 3) + CA_X_(r, 0) * U(2, 3) + CJ_X_(r, 0) * U(3, 3)

        CL_errj0_(r, 0) = (CL_X_(r, 0) - CL_X0(r, 0))
        CV_errj0_(r, 0) = (CV_X_(r, 0) - CV_X0(r, 0))
        CA_errj0_(r, 0) = (CA_X_(r, 0) - CA_X0(r, 0))
        CJ_errj0_(r, 0) = (CJ_X_(r, 0) - CJ_X0(r, 0))
    Next r

    CL_errj0 = Mat.Multiply(CL_errj0_, Mat.
Transpose(CL_errj0_))
    CV_errj0 = Mat.Multiply(CV_errj0_, Mat.
Transpose(CV_errj0_))
    CA_errj0 = Mat.Multiply(CA_errj0_, Mat.
Transpose(CA_errj0_))
    CJ_errj0 = Mat.Multiply(CJ_errj0_, Mat.
Transpose(CJ_errj0_))

    For r = 0 To 7
        For col = 0 To 7

            CL_P0(r, col) = (CL_P(r, col) + CL_errj0(r,
col)) * U(0, 0) + (CV_P(r, col) + CV_errj0(r, col)) *
U(1, 0) + (CA_P(r, col) + CA_errj0(r, col)) * U(2, 0) +
(CJ_P(r, col) + CJ_errj0(r, col)) * U(3, 0)
            CV_P0(r, col) = (CL_P(r, col) + CL_errj0(r,
col)) * U(0, 1) + (CV_P(r, col) + CV_errj0(r, col)) *
U(1, 1) + (CA_P(r, col) + CA_errj0(r, col)) * U(2, 1) +
(CJ_P(r, col) + CJ_errj0(r, col)) * U(3, 1)
            CA_P0(r, col) = (CL_P(r, col) + CL_errj0(r,
col)) * U(0, 2) + (CV_P(r, col) + CV_errj0(r, col)) *
U(1, 2) + (CA_P(r, col) + CA_errj0(r, col)) * U(2, 2) +
(CJ_P(r, col) + CJ_errj0(r, col)) * U(3, 2)
            CJ_P0(r, col) = (CL_P(r, col) + CL_errj0(r,
col)) * U(0, 3) + (CV_P(r, col) + CV_errj0(r, col)) *
U(1, 3) + (CA_P(r, col) + CA_errj0(r, col)) * U(2, 3) +
(CJ_P(r, col) + CJ_errj0(r, col)) * U(3, 3)
        Next col
    Next r

    'Updating value to KF parameters calculated in
Correct Step to be used in Predict Step
    For r = 0 To 7

        CL_X_(r, 0) = CL_X0(r, 0)
        CV_X_(r, 0) = CV_X0(r, 0)
        CA_X_(r, 0) = CA_X0(r, 0)
        CJ_X_(r, 0) = CJ_X0(r, 0)
```

```
        For col = 0 To 7
            CL_P(r, col) = CL_P0(r, col)
            CV_P(r, col) = CV_P0(r, col)
            CA_P(r, col) = CA_P0(r, col)
            CJ_P(r, col) = CJ_P0(r, col)
        Next col

    Next r

End Sub

Sub MM_filter_part2()
    Dim c  As Double
    Dim U(3)  As Double
    Dim C_X(7, 0) As Double
    Dim CL_errj() As Double, CV_errj() As Double, CA_
errj() As Double, CJ_errj() As Double
    Dim CL_errj_(7, 0) As Double, CV_errj_(7, 0) As
Double, CA_errj_(7, 0) As Double, CJ_errj_(7, 0) As
Double
    Dim CL_errj2() As Double, CV_errj2() As Double, CA_
errj2() As Double, CJ_errj2() As Double
    Dim CL_errj2_(7, 0) As Double, CV_errj2_(7, 0) As
Double, CA_errj2_(7, 0) As Double, CJ_errj2_(7, 0) As
Double
    Dim C_P(7, 7) As Double
    Dim CL_P2_(7, 7) As Double, CV_P2_(7, 7) As Double,
CA_P2_(7, 7) As Double, CJ_P2_(7, 7) As Double

    '--- IMM Step 3 --- Mode matched filtering
    'Likelihood funcion for each of the EKF
    MM0_V = Mat.Subtract(Z, CL_h_)
    MM0_VT = Mat.Transpose(MM0_V)
    MM0_S = Mat.Add(CL_HP2HT, CL_R)
    MM0_IS = Mat.Inv(MM0_S)
    MM0_S2 = Math.Sqr(Mat.Det(MM0_S))
    MM0_X = Mat.Multiply(Mat.Multiply(MM0_VT, MM0_IS),
MM0_V)
    MM0_X2 = MM0_X(0, 0)
    MM0_m = filters 'number of filters
    MM0_f = (1 / (((2 * 3.14) ^ (MM0_m / 2)) * MM0_S2)) ^
((-1 / 2) * MM0_X2)

    MM1_V = Mat.Subtract(Z, CV_h_)
    MM1_VT = Mat.Transpose(MM1_V)
    MM1_S = Mat.Add(CV_HP2HT, CV_R)
    MM1_IS = Mat.Inv(MM1_S)
    MM1_S2 = Math.Sqr(Mat.Det(MM1_S))
```

```
    MM1_X = Mat.Multiply(Mat.Multiply(MM1_VT, MM1_IS),
MM1_V)
    MM1_X2 = MM1_X(0, 0)
    MM1_m = filters 'number of filters
    MM1_f = (1 / (((2 * 3.14) ^ (MM1_m / 2)) * MM1_S2)) ^
((-1 / 2) * MM1_X2)

    MM2_V = Mat.Subtract(Z, CA_h_)
    MM2_VT = Mat.Transpose(MM2_V)
    MM2_S = Mat.Add(CA_HP2HT, CA_R)
    MM2_IS = Mat.Inv(MM2_S)
    MM2_S2 = Math.Sqr(Mat.Det(MM2_S))
    MM2_X = Mat.Multiply(Mat.Multiply(MM2_VT, MM2_IS),
MM2_V)
    MM2_X2 = MM2_X(0, 0)
    MM2_m = filters 'number of filters
    MM2_f = (1 / (((2 * 3.14) ^ (MM2_m / 2)) * MM2_S2)) ^
((-1 / 2) * MM2_X2)

    MM3_V = Mat.Subtract(Z, CJ_h_)
    MM3_VT = Mat.Transpose(MM3_V)
    MM3_S = Mat.Add(CJ_HP2HT, CJ_R)
    MM3_IS = Mat.Inv(MM3_S)
    MM3_S2 = Math.Sqr(Mat.Det(MM3_S))
    MM3_X = Mat.Multiply(Mat.Multiply(MM3_VT, MM3_IS),
MM3_V)
    MM3_X2 = MM3_X(0, 0)
    MM3_m = filters 'number of filters
    MM3_f = (1 / (((2 * 3.14) ^ (MM3_m / 2)) * MM3_S2)) ^
((-1 / 2) * MM3_X2)

    '--- IMM Step 4 --- Mode probability update

    c = MM0_f * cb(0) + MM1_f * cb(1) + MM2_f * cb(2) +
MM3_f * cb(3)

    U1(0) = (1 / c) * MM0_f * cb(0)
    U1(1) = (1 / c) * MM1_f * cb(1)
    U1(2) = (1 / c) * MM2_f * cb(2)
    U1(3) = (1 / c) * MM3_f * cb(3)

    'Preventing it going to zero
    For ww = 0 To 3
            If (U1(ww) <= 0) Then
                U1(ww) = 0.0000000001
            End If
    Next ww

    '--- IMM Step 5 --- For OUTPUT purposes only
(not part of algorithm recursions)
```

```
    For r = 0 To 7
        C_X(r, 0) = CL_h_(r, 0) * U1(0) + CV_h_(r, 0) *
U1(1) + CA_h_(r, 0) * U1(2) + CJ_h_(r, 0) * U1(3)

        CL_errj_(r, 0) = CL_h_(r, 0) - C_X(r, 0)
        CV_errj_(r, 0) = CV_h_(r, 0) - C_X(r, 0)
        CA_errj_(r, 0) = CA_h_(r, 0) - C_X(r, 0)
        CJ_errj_(r, 0) = CJ_h_(r, 0) - C_X(r, 0)
    Next r

    CL_errj = Mat.Multiply(CL_errj_, Mat.
Transpose(CL_errj_))
    CV_errj = Mat.Multiply(CV_errj_, Mat.
Transpose(CV_errj_))
    CA_errj = Mat.Multiply(CA_errj_, Mat.
Transpose(CA_errj_))
    CJ_errj = Mat.Multiply(CJ_errj_, Mat.
Transpose(CJ_errj_))

    For r = 0 To 7

        For col = 0 To 7
    C_P(r, col) = (CL_P2(r, col) + CL_errj(r, col)) *
U1(0) + (CV_P2(r, col) + CV_errj(r, col)) * U1(1) + (CA_
P2(r, col) + CA_errj(r, col)) * U1(2) + (CJ_P2(r, col) +
CJ_errj(r, col)) * U1(3)
        Next col
    Next r

    If (loop_cnt > PERIOD * 5) Then
If SNAP2ROAD Then

        Call Snap_to_Road_2.Start(Kalman_Filters.Convert_
Rad2Deg(C_X(0, 0) - offset_lat), Kalman_Filters.Convert_
Rad2Deg(C_X(1, 0) - offset_lon), Kalman_Filters.
Convert_Rad2Deg(Z(0, 0)), Kalman_Filters.Convert_
Rad2Deg(Z(1, 0)))
        C_X(0, 0) = Kalman_Filters.Convert_
Deg2Rad(curlong) + offset_lon
        C_X(1, 0) = Kalman_Filters.Convert_
Deg2Rad(curlat) + offset_lat

    '-----------------------------------------------------
    Set objPin = objMap.FindPushpin("Estimated
Location b")
    objPin.Delete
    estlong = Convert_Rad2Deg(C_X(0, 0))
    estlat = Convert_Rad2Deg(C_X(1, 0))
```

```
    Set objLoc = objMap.GetLocation(estlat, estlong)
    objMap.AddPushpin objLoc, "Estimated Location b"
    Set objCurLoc = objMap.FindPushpin("Estimated
Location b")
    objCurLoc.Symbol = 25
    '----------------------------------------------------

Else

    '----------------------------------------------------
    Set objPin = objMap.FindPushpin("Estimated Location")
    objPin.Delete
    estlong = Convert_Rad2Deg(C_X(0, 0))
    estlat = Convert_Rad2Deg(C_X(1, 0))
    Set objLoc = objMap.GetLocation(estlat, estlong)
    objMap.AddPushpin objLoc, "Estimated Location"
    Set objCurLoc = objMap.FindPushpin("Estimated
Location")
    objCurLoc.Symbol = 26
    '----------------------------------------------------

End If
    End If

    XM(0) = C_X(0, 0)    'x location
    XM(1) = C_X(1, 0)    'y location

End Sub

Sub CL_model_correct_loop()

   CL_HP2HT_loop = Mat.Multiply(Mat.Multiply(H, CL_P2_
loop), HT)
   CL_VRVT_loop = Mat.Multiply(Mat.Multiply(V, CL_R), VT)
   CL_S_loop = Mat.Add(CL_HP2HT_loop, CL_VRVT_loop)
   CL_Sinv_loop = Mat.Inv(CL_S_loop)

   CL_P2HT_loop = Mat.Multiply(CL_P2_loop, HT)
   CL_k_loop = Mat.Multiply(CL_P2HT_loop, CL_Sinv_loop)

   CL_k_loop = Dampen_K(CL_k_loop, "CL")

   CL_X_tmp_loop = Mat.Add(CL_h_loop_, Mat.Multiply(CL_k_
loop, Mat.Subtract(Z_loop, CL_h_loop_)))
   CL_P_loop = Mat.Multiply(Mat.Subtract(i, Mat.
Multiply(CL_k_loop, H)), CL_P2_loop)

   CL_X_loop(0, 0) = CL_X_tmp_loop(0, 0):    CL_X_loop(1,
0) = CL_X_tmp_loop(1, 0):    CL_X_loop(2, 0) = CL_X_tmp_
loop(2, 0):    CL_X_loop(3, 0) = CL_X_tmp_loop(3, 0)
```

```
    CL_X_loop(4, 0) = CL_X_tmp_loop(4, 0):    CL_X_loop(5,
0) = CL_X_tmp_loop(5, 0):    CL_X_loop(6, 0) = CL_X_tmp_
loop(6, 0):    CL_X_loop(7, 0) = CL_X_tmp_loop(7, 0)
    CL_X_loop_(0, 0) = CL_X_loop(0, 0):       CL_X_loop_
(1, 0) = CL_X_loop(1, 0):       CL_X_loop_(2, 0) = CL_X_
loop(2, 0):       CL_X_loop_(3, 0) = CL_X_loop(3, 0)
    CL_X_loop_(4, 0) = CL_X_loop(4, 0):       CL_X_loop_
(5, 0) = CL_X_loop(5, 0):       CL_X_loop_(6, 0) = CL_X_
loop(6, 0):       CL_X_loop_(7, 0) = CL_X_loop(7, 0)

End Sub

Sub CL_model_predict_loop()

    CL_X_loop(0, 0) = CL_X_loop_(0, 0)
    CL_X_loop(1, 0) = CL_X_loop_(1, 0)
    CL_X_loop(2, 0) = 0 'CL_X_loop_(2, 0)
    CL_X_loop(3, 0) = 0 'CL_X_loop_(3, 0)
    CL_X_loop(4, 0) = 0 'CL_X_loop_(4, 0)
    CL_X_loop(5, 0) = 0 'CL_X_loop_(5, 0)
    CL_X_loop(6, 0) = 0 'CL_X_loop_(6, 0)
    CL_X_loop(7, 0) = 0 'CL_X_loop_(7, 0)

    CL_APAT_loop = Mat.Multiply(Mat.Multiply(CL_A, CL_P_
loop), CL_AT)
    CL_WQWT_loop = Mat.Multiply(Mat.Multiply(CL_W, CL_Q),
CL_WT)
    CL_P2_loop = Mat.Add(CL_APAT_loop, CL_WQWT_loop)

    CL_h_loop_(0, 0) = CL_X_loop(0, 0): CL_h_loop_(1, 0) =
CL_X_loop(1, 0): CL_h_loop_(2, 0) = CL_X_loop(2, 0):
CL_h_loop_(3, 0) = CL_X_loop(3, 0)
    CL_h_loop_(4, 0) = CL_X_loop(4, 0): CL_h_loop_(5, 0) =
CL_X_loop(5, 0): CL_h_loop_(6, 0) = CL_X_loop(6, 0):
CL_h_loop_(7, 0) = CL_X_loop(7, 0)

End Sub

Sub CV_model_correct_loop()

    CV_HP2HT_loop = Mat.Multiply(Mat.Multiply(H, CV_P2_
loop), HT)
    CV_VRVT_loop = Mat.Multiply(Mat.Multiply(V, CV_R), VT)
    CV_S_loop = Mat.Add(CV_HP2HT_loop, CV_VRVT_loop)
    CV_Sinv_loop = Mat.Inv(CV_S_loop)

    CV_P2HT_loop = Mat.Multiply(CV_P2_loop, HT)
    CV_k_loop = Mat.Multiply(CV_P2HT_loop, CV_Sinv_loop)
```

```
   CV_k_loop = Dampen_K(CV_k_loop, "CV")
   CV_X_tmp_loop = Mat.Add(CV_h_loop_, Mat.Multiply(CV_k_
loop, Mat.Subtract(Z_loop, CV_h_loop_)))
   CV_P_loop = Mat.Multiply(Mat.Subtract(i, Mat.
Multiply(CV_k_loop, H)), CV_P2_loop)

   CV_X_loop(0, 0) = CV_X_tmp_loop(0, 0):    CV_X_loop(1,
0) = CV_X_tmp_loop(1, 0):    CV_X_loop(2, 0) = CV_X_tmp_
loop(2, 0):    CV_X_loop(3, 0) = CV_X_tmp_loop(3, 0)
   CV_X_loop(4, 0) = CV_X_tmp_loop(4, 0):    CV_X_loop(5,
0) = CV_X_tmp_loop(5, 0):    CV_X_loop(6, 0) = CV_X_tmp_
loop(6, 0):    CV_X_loop(7, 0) = CV_X_tmp_loop(7, 0)
   CV_X_loop_(0, 0) = CV_X_loop(0, 0):       CV_X_loop_
(1, 0) = CV_X_loop(1, 0):       CV_X_loop_(2, 0) = CV_X_
loop(2, 0):       CV_X_loop_(3, 0) = CV_X_loop(3, 0)
   CV_X_loop_(4, 0) = CV_X_loop(4, 0):       CV_X_loop_
(5, 0) = CV_X_loop(5, 0):       CV_X_loop_(6, 0) = CV_X_
loop(6, 0):       CV_X_loop_(7, 0) = CV_X_loop(7, 0)

End Sub

Sub CV_model_predict_loop()

   CV_X_loop(0, 0) = CV_X_loop_(0, 0) + CV_X_loop_(2, 0)
* delta_k_loop
   CV_X_loop(1, 0) = CV_X_loop_(1, 0) + CV_X_loop_(3, 0)
* delta_k_loop
   CV_X_loop(2, 0) = CV_X_loop_(2, 0)
   CV_X_loop(3, 0) = CV_X_loop_(3, 0)
   CV_X_loop(4, 0) = 0 'CV_X_loop_(4, 0)
   CV_X_loop(5, 0) = 0 'CV_X_loop_(5, 0)
   CV_X_loop(6, 0) = 0 'CV_X_loop_(6, 0)
   CV_X_loop(7, 0) = 0 'CV_X_loop_(7, 0)

   CV_APAT_loop = Mat.Multiply(Mat.Multiply(CV_A, CV_P_
loop), CV_AT)
   CV_WQWT_loop = Mat.Multiply(Mat.Multiply(CV_W, CV_Q),
CV_WT)
   CV_P2_loop = Mat.Add(CV_APAT_loop, CV_WQWT_loop)

   CV_h_loop_(0, 0) = CV_X_loop(0, 0): CV_h_loop_(1, 0) =
CV_X_loop(1, 0): CV_h_loop_(2, 0) = CV_X_loop(2, 0):
CV_h_loop_(3, 0) = CV_X_loop(3, 0)
   CV_h_loop_(4, 0) = CV_X_loop(4, 0): CV_h_loop_(5, 0) =
CV_X_loop(5, 0): CV_h_loop_(6, 0) = CV_X_loop(6, 0):
CV_h_loop_(7, 0) = CV_X_loop(7, 0)

End Sub

Sub CA_model_correct_loop()
```

```
   CA_HP2HT_loop = Mat.Multiply(Mat.Multiply(H, CA_P2_
loop), HT)
   CA_VRVT_loop = Mat.Multiply(Mat.Multiply(V, CA_R), VT)
   CA_S_loop = Mat.Add(CA_HP2HT_loop, CA_VRVT_loop)
   CA_Sinv_loop = Mat.Inv(CA_S_loop)

   CA_P2HT_loop = Mat.Multiply(CA_P2_loop, HT)
   CA_k_loop = Mat.Multiply(CA_P2HT_loop, CA_Sinv_loop)

   CA_k_loop = Dampen_K(CA_k_loop, "CA")
   CA_X_tmp_loop = Mat.Add(CA_h_loop_, Mat.Multiply(CA_k_
loop, Mat.Subtract(Z_loop, CA_h_loop_)))
   CA_P_loop = Mat.Multiply(Mat.Subtract(i, Mat.
Multiply(CA_k_loop, H)), CA_P2_loop)

   CA_X_loop(0, 0) = CA_X_tmp_loop(0, 0):    CA_X_loop(1,
0) = CA_X_tmp_loop(1, 0):    CA_X_loop(2, 0) = CA_X_tmp_
loop(2, 0):    CA_X_loop(3, 0) = CA_X_tmp_loop(3, 0)
   CA_X_loop(4, 0) = CA_X_tmp_loop(4, 0):    CA_X_loop(5,
0) = CA_X_tmp_loop(5, 0):    CA_X_loop(6, 0) = CA_X_tmp_
loop(6, 0):    CA_X_loop(7, 0) = CA_X_tmp_loop(7, 0)
   CA_X_loop_(0, 0) = CA_X_loop(0, 0):       CA_X_loop_
(1, 0) = CA_X_loop(1, 0):       CA_X_loop_(2, 0) = CA_X_
loop(2, 0):       CA_X_loop_(3, 0) = CA_X_loop(3, 0)
   CA_X_loop_(4, 0) = CA_X_loop(4, 0):       CA_X_loop_
(5, 0) = CA_X_loop(5, 0):       CA_X_loop_(6, 0) = CA_X_
loop(6, 0):       CA_X_loop_(7, 0) = CA_X_loop(7, 0)

End Sub

Sub CA_model_predict_loop()

   CA_X_loop(0, 0) = CA_X_loop_(0, 0) + CA_X_loop_(2, 0)
* delta_k_loop + (1 / 2) * CA_X_loop_(4, 0) * delta_k_
loop * delta_k_loop
   CA_X_loop(1, 0) = CA_X_loop_(1, 0) + CA_X_loop_(3, 0)
* delta_k_loop + (1 / 2) * CA_X_loop_(5, 0) * delta_k_
loop * delta_k_loop
   CA_X_loop(2, 0) = CA_X_loop_(2, 0) + CA_X_loop_(4, 0)
* delta_k_loop
   CA_X_loop(3, 0) = CA_X_loop_(3, 0) + CA_X_loop_(5, 0)
* delta_k_loop
   CA_X_loop(4, 0) = CA_X_loop_(4, 0)
   CA_X_loop(5, 0) = CA_X_loop_(5, 0)
   CA_X_loop(6, 0) = 0 'CA_X_loop_(6, 0)
   CA_X_loop(7, 0) = 0 'CA_X_loop_(7, 0)

   CA_APAT_loop = Mat.Multiply(Mat.Multiply(CA_A, CA_P_
loop), CA_AT)
```

```
   CA_WQWT_loop = Mat.Multiply(Mat.Multiply(CA_W, CA_Q),
CA_WT)
   CA_P2_loop = Mat.Add(CA_APAT_loop, CA_WQWT_loop)

   CA_h_loop_(0, 0) = CA_X_loop(0, 0): CA_h_loop_(1, 0) =
CA_X_loop(1, 0): CA_h_loop_(2, 0) = CA_X_loop(2, 0):
CA_h_loop_(3, 0) = CA_X_loop(3, 0)
   CA_h_loop_(4, 0) = CA_X_loop(4, 0): CA_h_loop_(5, 0) =
CA_X_loop(5, 0): CA_h_loop_(6, 0) = CA_X_loop(6, 0):
CA_h_loop_(7, 0) = CA_X_loop(7, 0)

End Sub

Sub CJ_model_correct_loop()

   CJ_HP2HT_loop = Mat.Multiply(Mat.Multiply(H, CJ_P2_
loop), HT)
   CJ_VRVT_loop = Mat.Multiply(Mat.Multiply(V, CJ_R),
VT)
   CJ_S_loop = Mat.Add(CJ_HP2HT_loop, CJ_VRVT_loop)
   CJ_Sinv_loop = Mat.Inv(CJ_S_loop)

   CJ_P2HT_loop = Mat.Multiply(CJ_P2_loop, HT)
   CJ_k_loop = Mat.Multiply(CJ_P2HT_loop, CJ_Sinv_loop)

   CJ_k_loop = Dampen_K(CJ_k_loop, "CJ")
   CJ_X_tmp_loop = Mat.Add(CJ_h_loop_, Mat.Multiply(CJ_k_
loop, Mat.Subtract(Z_loop, CJ_h_loop_)))
   CJ_P_loop = Mat.Multiply(Mat.Subtract(i, Mat.
Multiply(CJ_k_loop, H)), CJ_P2_loop)

   CJ_X_loop(0, 0) = CJ_X_tmp_loop(0, 0):    CJ_X_
loop(1, 0) = CJ_X_tmp_loop(1, 0):    CJ_X_loop(2, 0) =
CJ_X_tmp_loop(2, 0):    CJ_X_loop(3, 0) = CJ_X_tmp_
loop(3, 0)
   CJ_X_loop(4, 0) = CJ_X_tmp_loop(4, 0):    CJ_X_loop(5,
0) = CJ_X_tmp_loop(5, 0):    CJ_X_loop(6, 0) = CJ_X_tmp_
loop(6, 0):    CJ_X_loop(7, 0) = CJ_X_tmp_loop(7, 0)
   CJ_X_loop_(0, 0) = CJ_X_loop(0, 0):        CJ_X_loop_
(1, 0) = CJ_X_loop(1, 0):        CJ_X_loop_(2, 0) = CJ_X_
loop(2, 0):        CJ_X_loop_(3, 0) = CJ_X_loop(3, 0)
   CJ_X_loop_(4, 0) = CJ_X_loop(4, 0):        CJ_X_loop_
(5, 0) = CJ_X_loop(5, 0):        CJ_X_loop_(6, 0) = CJ_X_
loop(6, 0):        CJ_X_loop_(7, 0) = CJ_X_loop(7, 0)

End Sub

Sub CJ_model_predict_loop()
```

```
CJ_X_loop(0, 0) = CJ_X_loop_(0, 0) + CJ_X_loop_(2, 0) *
delta_k_loop + CJ_X_loop_(4, 0) * delta_k_loop * delta_k_
loop + CJ_X_loop_(6, 0) * delta_k_loop * delta_k_loop *
delta_k_loop
   CJ_X_loop(1, 0) = CJ_X_loop_(1, 0) + CJ_X_loop_(3, 0)
* delta_k_loop + CJ_X_loop_(5, 0) * delta_k_loop *
delta_k_loop + CJ_X_loop_(7, 0) * delta_k_loop * delta_k_
loop * delta_k_loop
   CJ_X_loop(2, 0) = CJ_X_loop_(2, 0) + CJ_X_loop_(4, 0)
* delta_k + (1 / 2) * CJ_X_loop_(6, 0) * delta_k *
delta_k
   CJ_X_loop(3, 0) = CJ_X_loop_(3, 0) + CJ_X_loop_(5, 0)
* delta_k + (1 / 2) * CJ_X_loop_(7, 0) * delta_k *
delta_k
   CJ_X_loop(4, 0) = CJ_X_loop_(4, 0) + CJ_X_loop_(6, 0)
* delta_k
   CJ_X_loop(5, 0) = CJ_X_loop_(5, 0) + CJ_X_loop_(7, 0)
* delta_k
   CJ_X_loop(6, 0) = CJ_X_loop_(6, 0)
   CJ_X_loop(7, 0) = CJ_X_loop_(7, 0)

CJ_APAT_loop = Mat.Multiply(Mat.Multiply(CJ_A, CJ_P_
loop), CJ_AT)
   CJ_WQWT_loop = Mat.Multiply(Mat.Multiply(CJ_W, CJ_Q),
CJ_WT)
   CJ_P2_loop = Mat.Add(CJ_APAT_loop, CJ_WQWT_loop)

   CJ_h_loop_(0, 0) = CJ_X_loop(0, 0): CJ_h_loop_(1, 0) =
CJ_X_loop(1, 0): CJ_h_loop_(2, 0) = CJ_X_loop(2, 0):
CJ_h_loop_(3, 0) = CJ_X_loop(3, 0)
   CJ_h_loop_(4, 0) = CJ_X_loop(4, 0): CJ_h_loop_(5, 0) =
CJ_X_loop(5, 0): CJ_h_loop_(6, 0) = CJ_X_loop(6, 0):
CJ_h_loop_(7, 0) = CJ_X_loop(7, 0)

End Sub

Sub MM_filter_part1_loop()
    Dim U(3, 3) As Double
    Dim CL_X0(7, 0) As Double, CV_X0(7, 0) As Double,
CA_X0(7, 0) As Double, CJ_X0(7, 0) As Double
    Dim CL_errj0_(7, 0) As Double, CV_errj0_(7, 0) As
Double, CA_errj0_(7, 0) As Double, CJ_errj0_(7, 0) As
Double
    Dim CL_errj0() As Double, CV_errj0() As Double, CA_
errj0() As Double, CJ_errj0() As Double
    Dim CL_P0(7, 7) As Double, CV_P0(7, 7) As Double,
CA_P0(7, 7) As Double, CJ_P0(7, 7) As Double
```

```
    '--- IMM Step 1 ---
    For col = 0 To 3

        cb_loop(col) = 0
        For row = 0 To 3
            cb_loop(col) = cb_loop(col) + BT(row, col) *
            U1_loop(row)
        Next row

        For row = 0 To 3
            U(row, col) = (1 / cb_loop(col)) * BT(row,
col) * U1_loop(row)

           'Preventing it going to zero
            If (U(row, col) <= 0) Then
                U(row, col) = 0.0000000001
            End If

        Next row

    Next col

    '--- IMM Step 2 ---
    For r = 0 To 7
        CL_X0(r, 0) = CL_X_loop_(r, 0) * U(0, 0) + CV_X_
loop_(r, 0) * U(1, 0) + CA_X_loop_(r, 0) * U(2, 0) +
CJ_X_loop_(r, 0) * U(3, 0)
        CV_X0(r, 0) = CL_X_loop_(r, 0) * U(0, 1) + CV_X_
loop_(r, 0) * U(1, 1) + CA_X_loop_(r, 0) * U(2, 1) +
CJ_X_loop_(r, 0) * U(3, 1)
        CA_X0(r, 0) = CL_X_loop_(r, 0) * U(0, 2) + CV_X_
loop_(r, 0) * U(1, 2) + CA_X_loop_(r, 0) * U(2, 2) +
CJ_X_loop_(r, 0) * U(3, 2)
        CJ_X0(r, 0) = CL_X_loop_(r, 0) * U(0, 3) + CV_X_
loop_(r, 0) * U(1, 3) + CA_X_loop_(r, 0) * U(2, 3) +
CJ_X_loop_(r, 0) * U(3, 3)

        CL_errj0_(r, 0) = CL_X_loop_(r, 0) - CL_X0(r, 0)
        CV_errj0_(r, 0) = CV_X_loop_(r, 0) - CV_X0(r, 0)
        CA_errj0_(r, 0) = CA_X_loop_(r, 0) - CA_X0(r, 0)
        CJ_errj0_(r, 0) = CJ_X_loop_(r, 0) - CJ_X0(r, 0)
    Next r

    CL_errj0 = Mat.Multiply(CL_errj0_, Mat.
Transpose(CL_errj0_))
    CV_errj0 = Mat.Multiply(CV_errj0_, Mat.
Transpose(CV_errj0_))
    CA_errj0 = Mat.Multiply(CA_errj0_, Mat.
Transpose(CA_errj0_))
```

```
    CJ_errj0 = Mat.Multiply(CJ_errj0_, Mat.
Transpose(CJ_errj0_))

    For r = 0 To 7
        For col = 0 To 7

            CL_P0(r, col) = (CL_P(r, col) + CL_errj0(r,
col)) * U(0, 0) + (CV_P(r, col) + CV_errj0(r, col)) *
U(1, 0) + (CA_P(r, col) + CA_errj0(r, col)) * U(2, 0) +
(CJ_P(r, col) + CJ_errj0(r, col)) * U(3, 0)
            CV_P0(r, col) = (CL_P(r, col) + CL_errj0(r,
col)) * U(0, 1) + (CV_P(r, col) + CV_errj0(r, col)) *
U(1, 1) + (CA_P(r, col) + CA_errj0(r, col)) * U(2, 1) +
(CJ_P(r, col) + CJ_errj0(r, col)) * U(3, 1)
            CA_P0(r, col) = (CL_P(r, col) + CL_errj0(r,
col)) * U(0, 2) + (CV_P(r, col) + CV_errj0(r, col)) *
U(1, 2) + (CA_P(r, col) + CA_errj0(r, col)) * U(2, 2) +
(CJ_P(r, col) + CJ_errj0(r, col)) * U(3, 2)
            CJ_P0(r, col) = (CL_P(r, col) + CL_errj0(r,
col)) * U(0, 3) + (CV_P(r, col) + CV_errj0(r, col)) *
U(1, 3) + (CA_P(r, col) + CA_errj0(r, col)) * U(2, 3) +
(CJ_P(r, col) + CJ_errj0(r, col)) * U(3, 3)

        Next col
    Next r

    'Updating value to KM parameters calculated in
Correct Step to be used in Predict Step
    For r = 0 To 7

        CL_X_loop_(r, 0) = CL_X0(r, 0)
        CV_X_loop_(r, 0) = CV_X0(r, 0)
        CA_X_loop_(r, 0) = CA_X0(r, 0)
        CJ_X_loop_(r, 0) = CJ_X0(r, 0)

        For col = 0 To 7
            CL_P_loop(r, col) = CL_P0(r, col)
            CV_P_loop(r, col) = CV_P0(r, col)
            CA_P_loop(r, col) = CA_P0(r, col)
            CJ_P_loop(r, col) = CJ_P0(r, col)
        Next col

    Next r

End Sub

Sub MM_filter_part2_loop()
```

```
    Dim c  As Double
    Dim U(3)  As Double
    Dim C_X(7, 0) As Double
    Dim CL_errj() As Double, CV_errj() As Double, CA_
errj() As Double, CJ_errj() As Double
    Dim CL_errj_(7, 0) As Double, CV_errj_(7, 0) As
Double, CA_errj_(7, 0) As Double, CJ_errj_(7, 0) As
Double
    Dim CL_errj2() As Double, CV_errj2() As Double, CA_
errj2() As Double, CJ_errj2() As Double
    Dim CL_errj2_(7, 0) As Double, CV_errj2_(7, 0) As
Double, CA_errj2_(7, 0) As Double, CJ_errj2_(7, 0) As
Double
    Dim C_P(7, 7) As Double

    MM0_V_loop = Mat.Subtract(Z_loop, CL_h_loop_)
    MM0_VT_loop = Mat.Transpose(MM0_V_loop)
    MM0_S_loop = Mat.Add(CL_HP2HT_loop, CL_R)
    MM0_IS_loop = Mat.Inv(MM0_S_loop)
    MM0_S2_loop = Math.Sqr(Mat.Det(MM0_S_loop))
    MM0_X_loop = Mat.Multiply(Mat.Multiply(MM0_VT_loop,
MM0_IS_loop), MM0_V_loop)
    MM0_X2_loop = MM0_X_loop(0, 0)
    MM0_m_loop = filters 'number of filters
    MM0_f_loop = (1 / (((2 * 3.14) ^ (MM0_m_loop / 2)) *
MM0_S2_loop)) ^ ((-1 / 2) * MM0_X2_loop)

    MM1_V_loop = Mat.Subtract(Z_loop, CV_h_loop_)
    MM1_VT_loop = Mat.Transpose(MM1_V_loop)
    MM1_S_loop = Mat.Add(CV_HP2HT_loop, CV_R)
    MM1_IS_loop = Mat.Inv(MM1_S_loop)
    MM1_S2_loop = Math.Sqr(Mat.Det(MM1_S_loop))
    MM1_X_loop = Mat.Multiply(Mat.Multiply(MM1_VT_loop,
MM1_IS_loop), MM1_V_loop)
    MM1_X2_loop = MM1_X_loop(0, 0)
    MM1_m_loop = filters 'number of filters
    MM1_f_loop = (1 / (((2 * 3.14) ^ (MM1_m_loop / 2)) *
    MM1_S2_loop)) ^ ((-1 / 2) * MM1_X2_loop)

    MM2_V_loop = Mat.Subtract(Z_loop, CA_h_loop_)
    MM2_VT_loop = Mat.Transpose(MM2_V_loop)
    MM2_S_loop = Mat.Add(CA_HP2HT_loop, CA_R)
    MM2_IS_loop = Mat.Inv(MM2_S_loop)
    MM2_S2_loop = Math.Sqr(Mat.Det(MM2_S_loop))
    MM2_X_loop = Mat.Multiply(Mat.Multiply(MM2_VT_loop,
MM2_IS_loop), MM2_V_loop)
    MM2_X2_loop = MM2_X_loop(0, 0)
    MM2_m_loop = filters 'number of filters
```

```
    MM2_f_loop = (1 / (((2 * 3.14) ^ (MM2_m_loop / 2)) *
    MM2_S2_loop)) ^ ((-1 / 2) * MM2_X2_loop)

    MM3_V_loop = Mat.Subtract(Z_loop, CJ_h_loop_)
    MM3_VT_loop = Mat.Transpose(MM3_V_loop)
    MM3_S_loop = Mat.Add(CJ_HP2HT_loop, CJ_R)
    MM3_IS_loop = Mat.Inv(MM3_S_loop)
    MM3_S2_loop = Math.Sqr(Mat.Det(MM3_S_loop))
    MM3_X_loop = Mat.Multiply(Mat.Multiply(MM3_VT_loop,
MM3_IS_loop), MM3_V_loop)
    MM3_X2_loop = MM3_X_loop(0, 0)
    MM3_m_loop = filters 'number of filters
    MM3_f_loop = (1 / (((2 * 3.14) ^ (MM3_m_loop / 2)) *
MM3_S2_loop)) ^ ((-1 / 2) * MM3_X2_loop)

    '--- IMM Step 4 ---

    c = MM0_f_loop * cb_loop(0) + MM1_f_loop * cb_loop(1)
+ MM2_f_loop * cb_loop(2) + MM3_f_loop * cb_loop(3)

    U1_loop(0) = (1 / c) * MM0_f_loop * cb_loop(0)
    U1_loop(1) = (1 / c) * MM1_f_loop * cb_loop(1)
    U1_loop(2) = (1 / c) * MM2_f_loop * cb_loop(2)
    U1_loop(3) = (1 / c) * MM3_f_loop * cb_loop(3)

    'Preventing it going to zero
    For ww = 0 To 3
            If (U1_loop(ww) <= 0) Then
                U1_loop(ww) = 0.0000000001
            End If
    Next ww

    '--- IMM Step 5 ---

    For r = 0 To 7
        C_X(r, 0) = CL_h_loop_(r, 0) * U1_loop(0) + CV_h_
        loop_(r, 0) * U1_loop(1) + CA_h_loop_(r, 0) *
        U1_loop(2) + CJ_h_loop_(r, 0) * U1_loop(3)

        CL_errj_(r, 0) = CL_h_loop_(r, 0) - C_X(r, 0)
        CV_errj_(r, 0) = CV_h_loop_(r, 0) - C_X(r, 0)
        CA_errj_(r, 0) = CA_h_loop_(r, 0) - C_X(r, 0)
        CJ_errj_(r, 0) = CJ_h_loop_(r, 0) - C_X(r, 0)
    Next r

    CL_errj = Mat.Multiply(CL_errj_, Mat.
Transpose(CL_errj_))
```

```
    CV_errj = Mat.Multiply(CV_errj_, Mat.
Transpose(CV_errj_))
    CA_errj = Mat.Multiply(CA_errj_, Mat.
Transpose(CA_errj_))
    CJ_errj = Mat.Multiply(CJ_errj_, Mat.
Transpose(CJ_errj_))

    For r = 0 To 7

        For col = 0 To 7
            C_P(r, col) = (CL_P2_loop(r, col) + CL_
errj(r, col)) * U1_loop(0) + (CV_P2_loop(r, col) + CV_
errj(r, col)) * U1_loop(1) + (CA_P2_loop(r, col) +
CA_errj(r, col)) * U1_loop(2) + (CJ_P2_loop(r, col) +
CJ_errj(r, col)) * U1_loop(3)

        Next col
    Next r

    If (loop_cnt > PERIOD * 5) Then

Var = "Estimated Location " & B_cnt
If SNAP2ROAD Then

        Call Snap_to_Road_2.Start(Kalman_Filters.
Convert_Rad2Deg(C_X(0, 0) - offset_lat), Kalman_Filters.
Convert_Rad2Deg(C_X(1, 0) - offset_lon), Kalman_Filters.
Convert_Rad2Deg(Z_loop(0, 0)), Kalman_Filters.Convert_
Rad2Deg(Z_loop(1, 0)))
        C_X(0, 0) = Kalman_Filters.Convert_
Deg2Rad(curlong) + offset_lon
        C_X(1, 0) = Kalman_Filters.Convert_
Deg2Rad(curlat) + offset_lat

    estlong = Convert_Rad2Deg(C_X(0, 0))
    estlat = Convert_Rad2Deg(C_X(1, 0))

Else

    estlong = Convert_Rad2Deg(C_X(0, 0))
    estlat = Convert_Rad2Deg(C_X(1, 0))

End If

    End If

    XM_loop(0) = C_X(0, 0)   'x location
    XM_loop(1) = C_X(1, 0)   'y location
```

```
End Sub

Sub Load_matrix_with_zeros(Mat() As Double, Rows As
Integer, cols As Integer)

  For l = 0 To Rows - 1
   For j = 0 To cols - 1
    Mat(l, j) = 0
   Next j
  Next l

End Sub

Sub Load_matrix_with_ones(Mat() As Double, Rows As
Integer, cols As Integer)

  For l = 0 To Rows - 1
   For j = 0 To cols - 1
    Mat(l, j) = 1
   Next j
  Next l

End Sub
```

A.4 MATLAB Code Used in Chapters 2 and 3

MATLAB Function	Description
sf_main	Starting point where system configuration variables are defined and KF/IMM/MMAE are called from as it looks through all log measurements
load_data	Used to read all the measurements from all the sensors and convert them to the same units
ekf_initialize	Used to initialize all the KF variables to be used in the correct/predict steps
ekf_correct	KF correct step mathematical calculations
ekf_predict	KF predict step mathematical calculations
ekf_models	KF models defined for system
ekf_update	Used to update the Q matrix of the KF for the DRWDE
imm_part1	Used for the first two steps of the IMM: calculation of the mixing probabilities and mixing
imm_part2	Used for the the next two steps of the IMM: mode-matched filtering and mode probability update
correction	Used when a correction method is enabled
format_results2	Used to properly format the estimated positions output

The **sf_main** function is the starting point to run this system. It first reads in the configuration variables used for the run. It has been coded to be very flexible to allow different numbers of KF, correction methods used, data points to use, measurements used, and a variety of more options that made evaluation of different setups easy to implement. The next section deals with initializing some of the variables used throughout the code, and then loads the first record from the log file into the array. At this point, the function goes into a loop to run through all data points specified; initially, it only runs the KF predict step for a few iterations to make sure all necessary data is available in the matrices before it moves into the normal system execution when all KF and IMM steps are executed. In the end, it adds all the results into three OUT matrices where the predicted locations are aligned to the actual GPS measurements to easily calculate the prediction errors.

```
function [OUT_val,OUT_err,OUT_data]=sf_main(data)

%  s* indicates available systems: s1 (sensor1=GPS), s2
(sensor2=ScanTool), s3 (sensor3=Accelerometer), sf
(sensorfusion=overall system)
%  s*_ekf are lists of EKF filters (each number
represents an EKF id supported in that system)
%  s*_P{#}, s*_W{#}, s*_Q{#}, s*_A{#}, s*_X{#} are arrays
of matrices where "#" indicates EKF id values the matrix
is for, and s* indicates for what system
%  s*_H, s*_V, s*_I, s*_R, s*_U1, s*_BT, s*_Z are
matrices shared between all the EKFs in each system (no #
needed)

Dk_orig = 0.1;            %in seconds (0.1 for 10Hz) NOTE:
anything less than 1 requires IMM to be running
sensors = [1 2 3];        %define which sensors to use in
the system   options=1,2,3
ekfs = [1 2 3];           %number of KFs in use (can NOT
change this without affecting BT)
use_ekf = 1;              %set to 0 for estimation of Z
only, or set to 1 to run system
use_imm = 1;              %set to 0 for EKF only run (no
IMM), or set to 1 for IMM run as well
est_sec_ahead = 3;        %set to far estimation location
3 seconds ahead (must also set use_imm=1), OTHERWISE set
to 0
est_sec_toGPS = 0;        %0=always estimate $est_sec_
ahead; 1=adjust estimation to always match record with
GPS value
use_rolling_window = 0;  %1=use avg rolling window in
load_data
```

```
use_rect_method = 0;      %1=use method of regression line
use_geom_method = 0;      %1=use method of geometric line

use_Q_calc_vars = 1;      %1=calculated Q variable;
0=simple Q variables
calc_missing_values = 1; %1=calc missing values based on
online sensors;  0=use IMM estimated values

gps_degORmtr = 2;         %1=degrees, 2=meters (for the
location units)
gps_difORtot = 1;         %1=diff between starting point
and current value, 2=full value (for the location units)
acc_vltORmtr = 1;         %1=volts, 2=meters/s^2
use_rectDreg = 0;         %0=no, 1=yes  (use recta de
regresion for acceleration values)
s_loop_start = 10;        %number of loop_count (rows of
data) to start system on [sensors section] (must be >3)

%--------------------
%Initializing variables that will hold the data from the
sensors for the different seconds
for numsecs=1:4
   gps_lat(numsecs) = 0;
   gps_lon(numsecs) = 0;
   gps_dir(numsecs) = 0;
   gps_vel(numsecs) = 0;
   sct_vel(numsecs) = 0;
   acc_avx(numsecs) = 0;
   acc_avy(numsecs) = 0;
   time(numsecs) = 0;
end
OUT(1,1)=0;
U(6,6)=0;
BT(6,6)=0;
cb(6,1)=0;
mm_f(6,1)=0;
%dir=0;
loop_count = 1;
start_time = 0;
Dk=Dk_orig;  %Dk will contain the gap between each set of
data (can be >0.1 when no data is available from any
sensor)
Dk2=Dk_orig;
Dks1=0;%Dks1 will contain the gap between each set of
data for s1
Dks2=0;%Dks2 will contain the gap between each set of
data for s2
```

```
Dks3=0;%Dks3 will contain the gap between each set of
data for s3
aborted=0;
sensor_status=0;
dif_an = 0;
ang1=0;
ang2=0;
ss0=0;
%--------------------

%Record starting coordinates to track distance from start
point
if gps_degORmtr == 1
   gps_lat_orig = data(1,4); %in degrees (Y direction)
   gps_lon_orig = data(1,5); %in degrees (X direction)
   gps_lat_orig2 = gps_lat_orig * 69.1 * 1609.344;
   gps_lon_orig2 = gps_lon_orig * 69.1703234283616 *
   cos(gps_lat_orig*0.0174532925199433) * 1609.344;
else
   gps_lon_orig = data(1,6); %in meters (X direction)
   gps_lat_orig = data(1,7); %in meters (Y direction)
   gps_lat_orig2 = gps_lat_orig;
   gps_lon_orig2 = gps_lon_orig;
end
if gps_difORtot == 1
   gps_lat_orig2 = 0;
   gps_lon_orig2 = 0;
end

%EKF initialize step (defines all corresponding variables
for all EKFs per sensor in use)
[H,I,A,P,BT,U1,U]=ekf_initialize(sensors,ekfs,Dk_orig);

%Loop through records in data array
[rows,cols] = size(data);
tot_loops=100000;
tot_recs=rows-20; %allow some extra rows after tot_recs
in case some sensors are needed for the last loop and
system needs to go past tot_recs

%Reading first record
[sensor_status,gps_lat2,gps_lon2,gps_dir2,gps_vel2,sct_
vel2,acc_avx2,acc_avy2,rec,Dk,Dk_prev]=load_data(Dk,Dk_
orig,sensors,data,rec,tot_recs,gps_lat_orig,gps_lon_
orig,gps_degORmtr,gps_difORtot,acc_vltORmtr,use_rolling_
window,loop_count,s_loop_start);

Dk_prev=Dk;
loop_count = loop_count+1;
```

```
cur_sec = data(rec,3)*1;
cur_sec_prev = cur_sec;
if rec > 1
   prev_sec = data(rec-1,3)*1;
end
next_sec = data(rec+1,3)*1;
cur_min = data(rec,2)*1;
cur_hr  = data(rec,1)*1;

%stop if first record does not have all sensors online
if ( sensor_status(1,1) == 0 | sensor_status(2,1) == 0 |
sensor_status(3,1) == 0 )
    disp(['Not all sensors online in first loop on record
' num2str(rec)]);
    aborted=1;
else
    %Saving current measurement into array
    gps_lat(1) = double(gps_lat2); %Latitude  = Y
direction (N/S)
    gps_lon(1) = double(gps_lon2); %Longitude = X
direction (E/W)
    gps_dir(1) = double(gps_dir2);
    gps_vel(1) = double(gps_vel2);
    sct_vel(1) = double(sct_vel2);
    acc_avx(1) = double(acc_avx2);
    acc_avy(1) = double(acc_avy2);
    time(1)    = cur_sec;
    %initialize some variables
    X_imm(1,1)= gps_lon2;
    X_imm(2,1)= gps_lat2;
    X_imm(3,1)= sct_vel2*sin(gps_dir2); % Vx
    X_imm(4,1)= sct_vel2*cos(gps_dir2); % Vy
    X_imm(5,1)= acc_avx2;
    X_imm(6,1)= acc_avy2;
    Z=double(X_imm);
    Z0=Z;
    Z_prev=Z;
    Z_prev2=Z;
    Z_prev3=Z;
    est0=Z;
    est=Z;
    Ap=0;
    Cp=0;

    OUT_data(loop_count,1)=rec;
    OUT_data(loop_count,2)=cur_hr;
    OUT_data(loop_count,3)=cur_min;
    OUT_data(loop_count,4)=cur_sec;
```

```
        OUT_data(loop_count,5)=Dk;
        OUT_data(loop_count,6)=Dk2;
        OUT_data(loop_count,7)=sensor_status(1,1);
        OUT_data(loop_count,8)=sensor_status(2,1);
        OUT_data(loop_count,9)=sensor_status(3,1);
        OUT_data(loop_count,10)=Z(1,1)-gps_lon_orig2;
        OUT_data(loop_count,11)=Z(2,1)-gps_lat_orig2;
        OUT_data(loop_count,12)=gps_dir(1)*57.2957795;
        OUT_data(loop_count,13)=sqrt(Z(3,1)^2+Z(4,1)^2);
        OUT_data(loop_count,14)=Z(5,1);
        OUT_data(loop_count,15)=Z(6,1);

    while loop_count <= tot_loops & rec <= tot_recs

        %Loading new data into variables
        Dk_prev_prev = Dk_prev;
        rec_prev = rec;
        sensor_status_prev = sensor_status;

[sensor_status,gps_lat2,gps_lon2,gps_dir2,gps_vel2,sct_
vel2,acc_avx2,acc_avy2,rec,Dk,Dk_prev]=load_data(Dk,Dk_
orig,sensors,data,rec,tot_recs,gps_lat_orig,gps_lon_
orig,gps_degORmtr,gps_difORtot,acc_vltORmtr,use_rolling_
window,loop_count,s_loop_start);

        %Calculating some variables
        cur_sec_prev = cur_sec;
        cur_min_prev = cur_min;
        cur_hr_prev  = cur_hr;
        cur_sec = data(rec,3)*1;
        cur_min = data(rec,2)*1;
        cur_hr  = data(rec,1)*1;
        next_sec = data(rec+1,3)*1;
        prev_sec = data(rec-1,3)*1;

        %stop if not enough data available to run the system
        if sensor_status==-1
            disp(['No sensors online in record '
            num2str(rec)]);
            aborted=1;
            break;
        end

        %Update Dk for each of the sensors to keep track of
time since last time it was online
        if sensor_status(1,1) == 0
           Dks1 = Dks1 + Dk;
        else
```

```
        Dks1 = 0;
     end
     if sensor_status(2,1) == 0
        Dks2 = Dks2 + Dk;
     else
        Dks2 = 0;
     end
     if sensor_status(3,1) == 0
        Dks3 = Dks3 + Dk;
     else
        Dks3 = 0;
     end

     %Keeping track of previous 3 measurements
     for t=2:4
        numsecs = 6-t;
        gps_lat(numsecs)  = double(gps_lat(numsecs-1));
        gps_lon(numsecs)  = double(gps_lon(numsecs-1));
        gps_dir(numsecs)  = double(gps_dir(numsecs-1));
        gps_vel(numsecs)  = double(gps_vel(numsecs-1));
        sct_vel(numsecs)  = double(sct_vel(numsecs-1));
        acc_avx(numsecs)  = double(acc_avx(numsecs-1));
        acc_avy(numsecs)  = double(acc_avy(numsecs-1));
        time(numsecs)     = time(numsecs-1);
     end
     %Saving current measurement into array
     gps_lat(1) = double(gps_lat2);
     gps_lon(1) = double(gps_lon2);
     gps_dir(1) = double(gps_dir2);
     gps_vel(1) = double(gps_vel2);
     sct_vel(1) = double(sct_vel2);
     acc_avx(1) = double(acc_avx2);
     acc_avy(1) = double(acc_avy2);
     time(numsecs)    = cur_sec;

     Z_prev3=double(Z_prev2); %saves last Z_prev2 matrix
 before loading new data into it
     Z_prev2=double(Z_prev);  %saves last Z_prev matrix
 before loading new data into it
     Z_prev=double(Z);  %saves last Z matrix before
 loading new data into it

     %LOAD Z matrices for each sensor
     %assumes each EKF for the same sensor will have the
     same matrix size [6x1]

             %--- acceleration ---
             if sensor_status(3,1)==1 %if acc is online
 use measured data
```

```
                Z(5,1) = acc_avx(1);
                Z(6,1) = acc_avy(1);
                ax2 = Z_prev(5,1);
                ay2 = Z_prev(6,1);
                %determining vectors based on
acceleration
                Anx = ax2*cos(gps_dir(2));
                Atx = ay2*sin(gps_dir(2));
                Any = ax2*sin(gps_dir(2));
                Aty = ay2*cos(gps_dir(2));
                if Z_prev(6,1) > 0
                   Ax  = Atx + Anx;
                   Ay  = Aty - Any;
                else
                   Ax  = Atx - Anx;
                   Ay  = Aty + Any;
                end
                %Calculate new velocities based on new
accelerations
                Vx  = Z_prev(3,1) + Ax *Dk_prev;
                Vy  = Z_prev(4,1) + Ay *Dk_prev;
                %Calculate new positions based on new
accelerations
                Sx = Z_prev(1,1) + ( Z_prev(3,1) )*Dk_
prev + (1/2)*( Ax )*Dk_prev^2;
                Sy = Z_prev(2,1) + ( Z_prev(4,1) )*Dk_
prev + (1/2)*( Ay )*Dk_prev^2;
                if Sy > Z_prev(2,1)
                   if  Sx >= Z_prev(1,1)
                      gps_dir(1) = atan( (Sx-Z_prev(1,1))
/ (Sy-Z_prev(2,1)) ) ;
                   else
                      gps_dir(1) = (2*pi) + atan( (Sx-Z_
prev(1,1)) / (Sy-Z_prev(2,1)) ) ;
                   end
                elseif Sy < Z_prev(2,1)
                   gps_dir(1) = pi + atan( (Sx-Z_
prev(1,1)) / (Sy-Z_prev(2,1)) ) ;
                else
                   if  Sx > Z_prev(1,1)
                      gps_dir(1) = (pi/2);
                   else
                      gps_dir(1) = (3*pi/2);
                   end
                end
             else
                if use_imm == 1
```

```
                    Z(5,1) = X_imm(5,1);  %using
previously estimated value (we don't want to derive it
from location for now)
                    Z(6,1) = X_imm(6,1);  %using
previously estimated value (we don't want to derive it
from location for now)
                 else
                    Z(5,1) = Z_prev(5,1);  %using previous
value
                    Z(6,1) = Z_prev(6,1);  %using previous
value
                 end
                 gps_dir(1) = gps_dir(2);
              end

              %--- GPS is online ---
              if sensor_status(1,1)==1  %If GPS is online
use measured data
                 %location
                 Z(1,1)=gps_lon(1); %x
                 Z(2,1)=gps_lat(1); %y
                 gps_dir(1) = double(gps_dir2); %use
actual data if sensor is online and ignore angle
calculated when s3 is on
              else %If GPS is offline then use previously
estimated data to assume current location
                 %location
                 if calc_missing_values == 1 && sensor_
status(3,1) == 1
                    Z(1,1)=Sx; %x
                    Z(2,1)=Sy; %y
                 else
                    if use_imm == 1
                       Z(1,1)=X_imm(1,1); %x
                       Z(2,1)=X_imm(2,1); %y
                    else
                       Z(1,1)=Z_prev(1,1); %x
                       Z(2,1)=Z_prev(2,1); %y
                    end
                 end
              end

              %--- Velocity ---
              if sensor_status(1,1)==1  %if GPS is online
use measured data
                 if sensor_status(2,1)==1 & ( sct_vel(1) >
0 | sct_vel(1) < 0 )  %If ST is online use measured data
(ST measurement preferred over GPS)
```

```
                    Z(3,1) = sct_vel(1)*sin(gps_dir(1));
                    Z(4,1) = sct_vel(1)*cos(gps_dir(1));
                 else
                    Z(3,1) = Z_prev(3,1);
                    Z(4,1) = Z_prev(4,1);
                 end
              else
                 Z(3,1) = Vx;
                 Z(4,1) = Vy;
              end
%Determine if method needs to be used or not
if use_rect_method==1 && sensor_status(3,1)==1

   %Determine miliseconds
   sec_txt = num2str(cur_sec);
   k = strfind(sec_txt, '.');
   if k > 0
      ss0 = str2num(sec_txt(k:k+1));
   else
      ss0 = 0;
   end

   %If miliseconds=0
   if ss0==0
      %moving angle into 1st and 4th quadrants [-pi,pi]
      ang1 = ang2;
      ang2 = 1.570796327 - gps_dir(1);
      if ang2 > 1.570796327
         ang2 = ang2 - pi;
      elseif ang2 < -1.570796327
         ang2 = ang2 + pi;
      end
      %Calculate elements of line
      if ang2 < -0.1 || ang2 > 0.1
         Ap = tan(ang2);
      else
         if ang1 > 0
            Ap = -128;
         else
            Ap = 127;
         end
      end
      Cp = Z(2,1) - Z(1,1)*Ap;
      %Calculate diet to the recta
      [est] =
ekf_models(rec,1,3,sensor_status,1,Dk_orig,Z,Z_prev,gps_
dir,acc_avx,acc_avy,time);
```

```
        dist1 = abs( Ap*est(1,1) - est(2,1) + Cp ) /
sqrt(Ap*Ap+1);
        dif_an = 0;
        a_n2 = acc_avx(1);
        for i=-8:8
           a_n = 0.15*i + a_n2;
           if i<0 || i>0
              [est] =
ekf_models(rec,1,3,sensor_status,1,Dk_orig,Z,Z_prev,gps_
dir,acc_avx,acc_avy,time);
              dist2 = abs( Ap*est0(1,1) - est0(2,1) + Cp )
/ sqrt(Ap*Ap+1);
              if dist1 > dist2
                 dist1 = dist2;
                 dif_an = 0.1*i;
              end
           end
        end
        avy1 = Z(6,1); avx1 = Z(5,1);
        if acc_avy(1) > -999
           avy1 = acc_avy(1); avx1 = acc_avx(1);
        end
        avy2 = Z_prev(6,1); avx2 = Z_prev(5,1);
        if acc_avy(2) > -999
           avy2 = acc_avy(2); avx2 = acc_avx(2);
        end
        avy3 = Z_prev2(6,1); avx3 = Z_prev2(5,1);
        if acc_avy(3) > -999
           avy3 = acc_avy(3); avx3 = acc_avx(3);
        end
        Z(6,1) = ( avy1+avy2+avy3 )/3;
        Z(5,1) = ( avx1+avx2+avx3 )/3 + dif_an*(2/3);
        dif_an = dif_an * (4/5);

        Z0=Z; %saving values when second is whole

     else
        avy1 = Z(6,1); avx1 = Z(5,1);
        if acc_avy(1) > -999
           avy1 = acc_avy(1); avx1 = acc_avx(1);
        end
        avy2 = Z_prev(6,1); avx2 = Z_prev(5,1);
        if acc_avy(2) > -999
           avy2 = acc_avy(2); avx2 = acc_avx(2);
        end
        avy3 = Z_prev2(6,1); avx3 = Z_prev2(5,1);
        if acc_avy(3) > -999
           avy3 = acc_avy(3); avx3 = acc_avx(3);
```

```
        end
        Z(6,1) = ( avy1+avy2+avy3 )/3;
        Z(5,1) = ( avx1+avx2+avx3 )/3 + dif_an*(2/3);

    end

end

        OUT_data(loop_count,1)=rec;
        OUT_data(loop_count,2)=cur_hr;
        OUT_data(loop_count,3)=cur_min;
        OUT_data(loop_count,4)=cur_sec;
        OUT_data(loop_count,5)=Dk;
        OUT_data(loop_count,6)=Dk2;
        OUT_data(loop_count,7)=sensor_status(1,1);
        OUT_data(loop_count,8)=sensor_status(2,1);
        OUT_data(loop_count,9)=sensor_status(3,1);
        OUT_data(loop_count,10)=Z(1,1)-gps_lon_orig2;
        OUT_data(loop_count,11)=Z(2,1)-gps_lat_orig2;
        OUT_data(loop_count,12)=gps_dir(1)*57.2957795;
        OUT_data(loop_count,13)=sqrt(Z(3,1)^2+Z(4,1)^2);
        OUT_data(loop_count,14)=Z(5,1);
        OUT_data(loop_count,15)=Z(6,1);

    if loop_count < s_loop_start
          %----- Initialization stage for the system ---

          %INITIALIZE X matrix to Z matrix
          %Loop through each KF defined
          [rows,cols] = size(ekfs);
           n=cols;
           for f=1:n
              X{f}=double(Z);
           end

%%%%%%%%%%%%%%%%%%%%%%%%%%%%%%%%%%%%%%%%%%%%%%%%%%%%%%%%%%%%%%%%%%%%%%%%%%%%%%%%%%%%%%%%%%%%%%%%%%%%%%%%%%%%%%%%%%%%%%%%%%%%%%%%%%%%%%%%%%%%%%%%%%%%%%%%%%%%%%%%%%%%%%%%%%%%%%%%%%
%%%%%%%%%%%
          %KF prediction step for sensors

          [R,Q]=ekf_update(sensor_status,ekfs,Dk,Dks1,Dks2,
Dks3,use_Q_calc_vars,use_R_calc_vars); %updating R and Q
matrices to use the current Dk
          %Loop through each KF defined
          [rows,cols] = size(ekfs);
          n=cols;
          for f=1:n
              if use_ekf == 1
                  [ekf_P] =
ekf_predict(sensor_status,A{f},P{f},Q{f});
```

```
            P{f}=double(ekf_P);
         end
         [ekf_X] =
ekf_models(rec,1,ekfs(f),sensor_status,Dk,Dk_orig,Z,Z_
prev,gps_dir,acc_avx,acc_avy,time);
         X{f}=double(ekf_X);
         if (f == 1)
            t=0;
         elseif (f == 2)
            t=3;
         elseif (f == 3)
            t=6;
         end
         OUT_data(loop_count,17+t)=X{f}
(1,1)-gps_lon_orig2;
         OUT_data(loop_count,18+t)=X{f}
(2,1)-gps_lat_orig2;
      end

   else
      %----- Running stage for the system ---

      %KF correct step for sensors
      %Loop through each KF defined
      [rows,cols] = size(ekfs);
      n = cols;
      for f=1:n
         if use_ekf == 1
            [ekf_X,ekf_P]=ekf_correct(ekfs(f),H,P{f},
R{f},I,Z,X{f});
            P{f}=double(ekf_P);
            X{f}=double(ekf_X);
         end
      end

      %IMM_part1
      if use_imm == 1
         [X,P,cb,U]=imm_part1(sensor_status,sensors,
ekfs,X,P,BT,U1,U,cb);
      end

      %KF prediction step for sensors

      [R,Q]=ekf_update(sensor_status,ekfs,Dk,Dks1,Dks2,
Dks3,use_Q_calc_vars,use_R_calc_vars); %updating R and Q
matrices to use the current Dk
      %Loop through each KF defined
      [rows,cols] = size(ekfs);
```

```
        n = cols;
        for f=1:n
           if use_ekf == 1
              [ekf_P] = ekf_predict(sensor_status,A{f},
P{f},Q{f});
              P{f}=double(ekf_P);
           end
           [ekf_X] = ekf_models(rec,1,ekfs(f),
sensor_status,Dk,Dk_orig,Z,Z_prev,gps_dir,acc_avx,acc_
avy,time);
           X{f}=double(ekf_X);
           if (f == 1)
              t=0;
           elseif (f == 2)
              t=3;
           elseif (f == 3)
              t=6;
           end
           OUT_data(loop_count,17+t)=X{f}
(1,1)-gps_lon_orig2;
           OUT_data(loop_count,18+t)=X{f}
(2,1)-gps_lat_orig2;
        end

        %IMM_part2
        if use_imm == 1
           [U1,X_imm,mm_f] = imm_part2(sensor_status,sens
ors,ekfs,X,P,H,R,cb,mm_f);
        OUT_data(loop_count,26)=X_imm(1,1)-gps_lon_orig2;
        OUT_data(loop_count,27)=X_imm(2,1)-gps_lat_orig2;
        end

         %Estimating position 3 seconds ahead.
         if est_sec_ahead > 0

            %Get miliseconds for cur_sec
            if Dk_orig < 1 & est_sec_toGPS == 1
               sec_txt = num2str(cur_sec);
               k = strfind(sec_txt, '.');
               if k > 0
                  ss = str2num(sec_txt(k:k+1));
               else
                  ss = 0;
               end
               Dk2 = est_sec_ahead - ss;
               if ss > est_sec_ahead | Dk2 >
est_sec_ahead
```

```
            Dk2 = est_sec_ahead;
         end
      else
         Dk2 = est_sec_ahead;
      end
      OUT_data(loop_count,6)=Dk2;

      if Dk_orig == 1 & est_mid_points == 1
         totloops = 9;  %going for 3.0 to 3.9
for 1sec estimation
      else
         totloops = 0;
      end
      for d=0:totloops
         Dk2 = est_sec_ahead + (d/10);

      %Run KF again but this time using a larger Dk
      [rows,cols] = size(ekfs);
      n = cols;
      for f=1:n
         [ekf_X_ahead] =
ekf_models(rec,1,ekfs(f),sensor_status,Dk2,Dk_orig,Z,Z_
prev,gps_dir,acc_avx,acc_avy,time);
         if ekfs(f)==3 && use_geom_method == 1
            [ekf_X_ahead,Ap,Cp] =
correction(ss0,ekf_X_ahead,est0,Ap,Cp,ang1,ang2,Z,Z0,rec,
cur_hr,cur_min,cur_sec);
            est0 = ekf_X_ahead;
         end
         X_ahead{f}=ekf_X_ahead;
         if (f == 1)
            t=0;
         elseif (f == 2)
            t=3;
         elseif (f == 3)
            t=6;
         end
         OUT_data(loop_count,29+t)=X_ahead{f}
(1,1)-gps_lon_orig2;
         OUT_data(loop_count,30+t)=X_ahead{f}
(2,1)-gps_lat_orig2;
      end

      if use_imm == 1
         %Run IMM_part2 to merge the results from
the KF for this 3sec ahead estimation
         [U1_ahead,X_imm_ahead,mm_f_ahead] =
imm_part2(sensor_status,sensors,ekfs,X_
ahead,P,H,R,cb,mm_f);
```

```
               OUT_data(loop_count,38)=X_imm_ahead(1,1)
-gps_lon_orig2;
               OUT_data(loop_count,39)=X_imm_ahead(2,1)
-gps_lat_orig2;
            else
               X_ahead = X;
               X_imm_ahead = X_imm;
            end

            if d == 0
               %Record results
               [OUT,start_time] =
format_results2(data,sensor_status,sensors,ekfs,use_
imm,OUT,Z,X_ahead,X_imm_ahead,Dk2,Dk,Dk_orig,loop_
count,s_loop_start,rec,est_sec_ahead,start_time);
            end

            loop_count = loop_count+1;

         end
         loop_count = loop_count-1;

         else
            X_ahead = X;
            X_imm_ahead = X_imm;
           %Record results
           [OUT,start_time] =
format_results2(data,sensor_status,sensors,ekfs,use_
imm,OUT,Z,X_ahead,X_imm_ahead,Dk2,Dk,Dk_orig,loop_
count,s_loop_start,rec,est_sec_ahead,start_time);
         end

end

    loop_count = loop_count+1;

end

%Recording array with full values
OUT_val = OUT;

%Select type of data to be displayed
type=1;  %2=location, 1=error
if type == 1 & aborted == 0

   %Get total number of EKFs defined
   [rows,cols] = size(ekfs);
   total_ekfs=cols;
```

```
    %Loop through all rows in OUT
    [rows,cols] = size(OUT);
    row_count=2;
    SUM(1:2,1:50)=0; %setting the whole array to zero
    while row_count < rows

       sensors_online = OUT(row_count,7) + OUT(row_
count,8) + OUT(row_count,9);
       if sensors_online > 0 & OUT(row_count,1) >= 0

       %Location X
       n=13; %start from this column number (which
contains GPS latitude data)
       if OUT(row_count,n) ~= 0
       for c=1:total_ekfs  %loop through defined EKFs
          if OUT(row_count,n+c) ~= 0
             OUT(row_count,n+c)=abs( OUT(row_count,n+c)-
OUT(row_count,n) );
             SUM(1,n+c) = SUM(1,n+c) + OUT(row_count,n+c);
             SUM(2,n+c) = SUM(2,n+c) + 1;
          end
       end
       if use_imm == 1  %record IMM result on last column
of this group
          if OUT(row_count,n+1+c) ~= 0
             OUT(row_count,n+1+c)=abs( OUT(row_
count,n+1+c)-OUT(row_count,n) );
             SUM(1,n+1+c) = SUM(1,n+1+c) +
OUT(row_count,n+1+c);
             SUM(2,n+1+c) = SUM(2,n+1+c) + 1;
          end
       end
       end
       %Location Y
       n=n+1+total_ekfs+2; %leave one column empty in
between lat and lon sections (the column will have all
zeros)
       if OUT(row_count,n) ~= 0
       for c=1:total_ekfs  %loop through defined EKFs
          if OUT(row_count,n+c) ~= 0
             OUT(row_count,n+c)=abs( OUT(row_count,n+c)-
OUT(row_count,n) );
             SUM(1,n+c) = SUM(1,n+c) + OUT(row_count,n+c);
             SUM(2,n+c) = SUM(2,n+c) + 1;
          end
       end
       if use_imm == 1  %record IMM result on last column
of this group
```

```
         if OUT(row_count,n+1+c) ~= 0
            OUT(row_count,n+1+c)=abs( OUT(row_
count,n+1+c)-OUT(row_count,n) );
            SUM(1,n+1+c) = SUM(1,n+1+c) +
OUT(row_count,n+1+c);
            SUM(2,n+1+c) = SUM(2,n+1+c) + 1;
         end
      end
      end
      %Vel_x
      n=n+1+total_ekfs+2; %leave one column empty in
between lat and lon sections (the column will have all
zeros)
      if OUT(row_count,n) ~= 0
      for c=1:total_ekfs  %loop through defined EKFs
         if OUT(row_count,n+c) ~= 0
            OUT(row_count,n+c)=abs( OUT(row_count,n+c)-
OUT(row_count,n) );
            SUM(1,n+c) = SUM(1,n+c) + OUT(row_count,n+c);
            SUM(2,n+c) = SUM(2,n+c) + 1;
         end
      end
      if use_imm == 1  %record IMM result on last column
of this group
         if OUT(row_count,n+1+c) ~= 0
            OUT(row_count,n+1+c)=abs( OUT(row_
count,n+1+c)-OUT(row_count,n) );
            SUM(1,n+1+c) = SUM(1,n+1+c) +
OUT(row_count,n+1+c);
            SUM(2,n+1+c) = SUM(2,n+1+c) + 1;
         end
      end
      end
      %Vel_y
      n=n+1+total_ekfs+2; %leave one column empty in
between lat and lon sections (the column will have all
zeros)
      if OUT(row_count,n) ~= 0
      for c=1:total_ekfs  %loop through defined EKFs
         if OUT(row_count,n+c) ~= 0
            OUT(row_count,n+c)=abs( OUT(row_count,n+c)-
OUT(row_count,n) );
            SUM(1,n+c) = SUM(1,n+c) + OUT(row_count,n+c);
            SUM(2,n+c) = SUM(2,n+c) + 1;
         end
      end
      if use_imm == 1  %record IMM result on last column
of this group
```

```
            if OUT(row_count,n+1+c) ~= 0
               OUT(row_count,n+1+c)=abs( OUT(row_
count,n+1+c)-OUT(row_count,n) );
               SUM(1,n+1+c) = SUM(1,n+1+c) +
OUT(row_count,n+1+c);
               SUM(2,n+1+c) = SUM(2,n+1+c) + 1;
            end
         end
         end
         %Acc_x
         n=n+1+total_ekfs+2; %leave one column empty in
between lat and lon sections (the column will have all
zeros)
         if OUT(row_count,n) ~= 0
         for c=1:total_ekfs  %loop through defined EKFs
            if OUT(row_count,n+c) ~= 0
               OUT(row_count,n+c)=abs( OUT(row_count,n+c)-
OUT(row_count,n) );
               SUM(1,n+c) = SUM(1,n+c) + OUT(row_count,n+c);
               SUM(2,n+c) = SUM(2,n+c) + 1;

            end
         end
         if use_imm == 1  %record IMM result on last column
of this group
            if OUT(row_count,n+1+c) ~= 0
               OUT(row_count,n+1+c)=abs( OUT(row_
count,n+1+c)-OUT(row_count,n) );
               SUM(1,n+1+c) = SUM(1,n+1+c) +
OUT(row_count,n+1+c);
               SUM(2,n+1+c) = SUM(2,n+1+c) + 1;
            end
         end
         end
         %Acc_y
         n=n+1+total_ekfs+2; %leave one column empty in
between lat and lon sections (the column will have all
zeros)
         if OUT(row_count,n) ~= 0
         for c=1:total_ekfs  %loop through defined EKFs
            if OUT(row_count,n+c) ~= 0
               OUT(row_count,n+c)=abs( OUT(row_count,n+c)-
OUT(row_count,n) );
               SUM(1,n+c) = SUM(1,n+c) + OUT(row_count,n+c);
               SUM(2,n+c) = SUM(2,n+c) + 1;
            end
         end
```

```
        if use_imm == 1  %record IMM result on last column
of this group
            if OUT(row_count,n+1+c) ~= 0
                OUT(row_count,n+1+c)=abs( OUT(row_
count,n+1+c)-OUT(row_count,n) );
                SUM(1,n+1+c) = SUM(1,n+1+c) +
OUT(row_count,n+1+c);
                SUM(2,n+1+c) = SUM(2,n+1+c) + 1;
            end
        end
        end

        end

        row_count = row_count+1;

    end

    OUT(row_count,1:n+1+total_ekfs)=0;
    row_count = row_count+1; %leave an empty row before
results

    %Calculating average error for each column
    %Location X
    n=13; %start from this column number (which contains
GPS latitude data)
    if SUM(1,n+c) > 0
    for c=1:total_ekfs  %loop through defined EKFs
        OUT(row_count,n+c)=SUM(1,n+c);
        OUT(row_count+1,n+c)=SUM(2,n+c);
        OUT(row_count+2,n+c)=SUM(1,n+c)/SUM(2,n+c);
    end
    if use_imm == 1  %record IMM result on last column of
this group
        OUT(row_count,n+1+c)=SUM(1,n+1+c);
        OUT(row_count+1,n+1+c)=SUM(2,n+1+c);
        OUT(row_count+2,n+1+c)=SUM(1,n+1+c)/SUM(2,n+1+c);
    end
    end
    %Location Y
    n=n+1+total_ekfs+2; %leave one column empty in between
lat and lon sections (the column will have all zeros)
    if SUM(1,n+c) > 0
    for c=1:total_ekfs  %loop through defined EKFs
        OUT(row_count,n+c)=SUM(1,n+c);
        OUT(row_count+1,n+c)=SUM(2,n+c);
        OUT(row_count+2,n+c)=SUM(1,n+c)/SUM(2,n+c);
```

```
   end
   if use_imm == 1  %record IMM result on last column of
this group
      OUT(row_count,n+1+c)=SUM(1,n+1+c);
      OUT(row_count+1,n+1+c)=SUM(2,n+1+c);
      OUT(row_count+2,n+1+c)=SUM(1,n+1+c)/SUM(2,n+1+c);
   end
   end
   %Vel_x
   n=n+1+total_ekfs+2; %leave one column empty in between
lat and lon sections (the column will have all zeros)
   if SUM(1,n+c) > 0
   for c=1:total_ekfs  %loop through defined EKFs
      OUT(row_count,n+c)=SUM(1,n+c);
      OUT(row_count+1,n+c)=SUM(2,n+c);
      OUT(row_count+2,n+c)=SUM(1,n+c)/SUM(2,n+c);
   end
   if use_imm == 1  %record IMM result on last column of
this group
      OUT(row_count,n+1+c)=SUM(1,n+1+c);
      OUT(row_count+1,n+1+c)=SUM(2,n+1+c);
      OUT(row_count+2,n+1+c)=SUM(1,n+1+c)/SUM(2,n+1+c);
   end
   end
   %Vel_y
   n=n+1+total_ekfs+2; %leave one column empty in between
lat and lon sections (the column will have all zeros)
   if SUM(1,n+c) > 0
   for c=1:total_ekfs  %loop through defined EKFs
      OUT(row_count,n+c)=SUM(1,n+c);
      OUT(row_count+1,n+c)=SUM(2,n+c);
      OUT(row_count+2,n+c)=SUM(1,n+c)/SUM(2,n+c);
   end
   if use_imm == 1  %record IMM result on last column of
this group
      OUT(row_count,n+1+c)=SUM(1,n+1+c);
      OUT(row_count+1,n+1+c)=SUM(2,n+1+c);
      OUT(row_count+2,n+1+c)=SUM(1,n+1+c)/SUM(2,n+1+c);
   end
   end
   %Acc_x
   n=n+1+total_ekfs+2; %leave one column empty in between
lat and lon sections (the column will have all zeros)
   if SUM(1,n+c) > 0
   for c=1:total_ekfs  %loop through defined EKFs
      OUT(row_count,n+c)=SUM(1,n+c);
      OUT(row_count+1,n+c)=SUM(2,n+c);
```

```
        OUT(row_count+2,n+c)=SUM(1,n+c)/SUM(2,n+c);
    end
    if use_imm == 1  %record IMM result on last column of
this group
        OUT(row_count,n+1+c)=SUM(1,n+1+c);
        OUT(row_count+1,n+1+c)=SUM(2,n+1+c);
        OUT(row_count+2,n+1+c)=SUM(1,n+1+c)/SUM(2,n+1+c);
    end
    end
    %Acc_y
    n=n+1+total_ekfs+2; %leave one column empty in between
lat and lon sections (the column will have all zeros)
    if SUM(1,n+c) > 0
    for c=1:total_ekfs  %loop through defined EKFs
        OUT(row_count,n+c)=SUM(1,n+c);
        OUT(row_count+1,n+c)=SUM(2,n+c);
        OUT(row_count+2,n+c)=SUM(1,n+c)/SUM(2,n+c);
    end
    if use_imm == 1  %record IMM result on last column of
this group
        OUT(row_count,n+1+c)=SUM(1,n+1+c);
        OUT(row_count+1,n+1+c)=SUM(2,n+1+c);
        OUT(row_count+2,n+1+c)=SUM(1,n+1+c)/SUM(2,n+1+c);
    end
    end

    %Recording array with error values
    OUT_err = OUT;

end  %---end of type == 1 & aborted == 0

end

end  %---end of main function
```

The **load_data** function is used to read all the measurements from all the sensors and then convert them to the same units so they can be used and compared easily throughout the system. It also identifies any missing measurements if a sensor was offline, so that the system can treat that measurement accordingly.

```
function
[sensor_status,gps_lat2,gps_lon2,gps_dir2,gps_vel2,sct_
vel2,acc_avx2,acc_avy2,rec,Dk,Dk_prev]=load_data(Dk,Dk_
orig,sensors,data,rec,tot_rows,gps_lat_orig,gps_lon_
```

```
orig,gps_degORmtr,gps_difORtot,acc_vltORmtr,use_rolling_
window,loop_count,s_loop_start)

loop=1;
rec_0=rec;
rec=rec+1;
if (rec-1) == 0
   Dk_prev = Dk;
end
while loop==1
   %Loading new data into variables
   if gps_degORmtr == 1
      gps_lat_0 = double(data(rec,4)); %Latitude in
degrees (Y direction)
      gps_lon_0 = double(data(rec,5)); %Longitude in
degrees (X direction)
   else
      gps_lon_0 = double(data(rec,6)); %Longitude in
meters (X direction)
      gps_lat_0 = double(data(rec,7)); %Latitude in
meters (Y direction)
   end
   gps_dir_0  = double(data(rec,8)); %in degrees
   gps_vel_0  = double(data(rec,9)); %in kph
   sct_vel_0  = double(data(rec,10)); %in mph
   acc_avx_0  = double(data(rec,11));  % acc_normal
   acc_avy_0  = double(data(rec,12));  % acc_tangencial

   %code to create a rolling average window on
accelerations
   if use_rolling_window == 1
      if rec > 2 & Dk_orig < 1
         avx_cnt=0;    avx_tot=0;
         avy_cnt=0;    avy_tot=0;
         for r=-2:0  %this only uses current and past 2
values
            if isempty( data(rec+r,11) ) == 0
               avx_tot = avx_tot + double(data(rec+r,11));
               avx_cnt  = avx_cnt+1;
            end
            if isempty( data(rec+r,12) ) == 0
               avy_tot = avy_tot + double(data(rec+r,12));
               avy_cnt  = avy_cnt+1;
            end
         end
         if avx_tot > 0
            acc_avx_0  = avx_tot / avx_cnt;  % acc_normal
         end
```

```
         if avy_tot > 0
            acc_avy_0  = avy_tot / avy_cnt;  % acc_normal
         end
      end
   end

   %Check is Sensor 1 is offline (1=online, 0=offline)
   sensor_status(1,1) = 1;
   %Adding exception because in a few instances all GPS 
values are present except for speed
   if isnan(gps_vel_0)
      gps_vel_0 = 0;
   end
   if isnan(gps_lat_0) | isnan(gps_lon_0) | isnan(gps_
dir_0) | isnan(gps_vel_0)
      sensor_status(1,1) = 0;
   end
   %Check is Sensor 2 is offline
   sensor_status(2,1) = 1;
   if isnan(sct_vel_0)
       sensor_status(2,1) = 0;
   end
   %Check is Sensor 3 is offline
   sensor_status(3,1) = 1;
   if isnan(acc_avx_0) | isnan(acc_avx_0)
       sensor_status(3,1) = 0;
   end
   %define which sensors are online
   s(1)=0; s(2)=0; s(3)=0;
   [rows,cols] = size(sensors);
   for t=1:cols
      s(sensors(t)) = sensor_status(sensors(t),1);
   end

   %Check that at least 1 sensor is online, otherwise 
read next record
   %check what fraction time to lookup data for (Dk=1 or 
Dk=0.1)
   if Dk_orig < 1 | s(1) == 1
      if Dk_orig < 1
         if (rec-1) > 0
            %Adjust correct Dk if data is not available 
for current time fraction
            cur_sec_0 = ( ( ( data(rec-1,1) * 60 ) + 
data(rec-1,2) )*60 ) + data(rec-1,3) ;
            cur_sec_1 = (( (data(rec,1)*60)+data(rec,2)
)*60)+data(rec,3);
```

```
            cur_sec_2 = (( (data(rec+1,1)*60)+data(
rec+1,2) )*60)+data(rec+1,3);
            Dk_prev = cur_sec_1 - cur_sec_0;
            Dk = cur_sec_2 - cur_sec_1;
            loop=0;  %stop loop
         else
             loop=0;  %Assumes rec=0 has all sensors
online
         end
      elseif s(1) == 1
         Dk_prev = Dk_orig;
         Dk = Dk_orig;
         loop=0;  %Assumes rec=0 has all sensors online
      end
   else
      rec = rec+1;
   end

end

if loop==0  % 0 means that there is enough data to run
the system

    %Converting units of each reading to have common
units
    if sensor_status(1,1) == 1
        if gps_degORmtr == 1
           if gps_difORtot == 1
              gps_lat2 = (gps_lat_0-gps_lat_orig) * 69.1
* 1609.344; %in meters (converted from degrees)
              gps_lon2 = (gps_lon_0-gps_lon_orig) *
69.1703234283616 * cos(gps_lat_0*0.0174532925199433) *
1609.344; %in meters (converted from degrees)
           else
              gps_lat2 = (gps_lat_0) * 69.1 * 1609.344;
%in meters (converted from degrees)
              gps_lon2 = (gps_lon_0) * 69.1703234283616 *
cos(gps_lat_0*0.0174532925199433) * 1609.344; %in meters
(converted from degrees)
           end
        else
           if gps_difORtot == 1
              gps_lat2 = gps_lat_0-gps_lat_orig; %in
meters
              gps_lon2 = gps_lon_0-gps_lon_orig; %in
meters
           else
```

```
                gps_lat2 = gps_lat_0; %in meters
                gps_lon2 = gps_lon_0; %in meters
            end
        end
        gps_dir2 = gps_dir_0*pi/180; %in radians
(converted from degrees)
        gps_vel2 = gps_vel_0*1000/3600; %in meters/sec
(converted from km per hour)
     else
        gps_lat2 = -999; %error
        gps_lon2 = -999; %error
        gps_dir2 = -999; %error
        gps_vel2 = -999; %error
    end
    if sensor_status(2,1) == 1
        sct_vel2 = sct_vel_0*1000/3600; %in meters/sec
(converted from km per hour)
    else
       sct_vel2 = -999; %error
    end
    if sensor_status(3,1) == 1
       if acc_vltORmtr == 1
          acc_avx2 = ((acc_avx_0-
2.45)*21.9368513*1609.344/3600) * 1.5; %in meters/sec^2
(converted from miles/hour*sec)
          acc_avy2 = (acc_avy_0-
2.45)*21.9368513*1609.344/3600; %in meters/sec^2
(converted from miles/hour*sec)
       else
          acc_avx2 = acc_avx_0b; %in meters/sec^2
          acc_avy2 = acc_avy_0b; %in meters/sec^2
       end
    else
       acc_avx2 = -999; %error
       acc_avy2 = -999; %error
    end

else
    %Not enough data available to run the system for the
given conditions (variables at the top of sf_main)
    sensor_status=-1;
    gps_lat2=-1;
    gps_lon2=-1;
    gps_dir2=-1;
    gps_vel2=-1;
    sct_vel2=-1;
    acc_avx2=-1;
```

```
        acc_avy2=-1;
    end

        return;
    end
```

The **ekf_initialize** is used to initialize all the KF variables to be used in the correct/predict steps.

```
function
[H,I,A,P,BT,U1,U]=ekf_initialize(sensors,ekfs,Dk)

sig_p = 0.01;
sig_m = 1;

%Loop through all EKFs to set their specific W,A,Q,P
arrays
[rows,cols] = size(ekfs);
n = cols;
for f=1:n
    %--- KF - A
    for z1=1:6
        for z2=1:6
            A{f}(z1,z2) = 0;
        end
        A{f}(z1,z1) = 1;
    end
    A{f}(1,2)=1*Dk;
    A{f}(2,3)=1*Dk;
    %--- KF - P
    for z1=1:6
        for z2=1:6
            P{f}(z1,z2) = 0;
        end
        P{f}(z1,z1) = 1;
    end
end

%Initialize variables that are the same for all KF
filters
%--- KF - H
for z1=1:6
    for z2=1:6
        H(z1,z2) = 0;
    end
```

```
    H(z1,z1) = 1;
  end
  %--- KF - I
  for z1=1:6
     for z2=1:6
        I(z1,z2) = 0;
     end
     I(z1,z1) = 1;
  end

  %IMM - U1 initialization
  U1(1,1) = 0.00001;
  U1(2,1) = 0.00001;
  U1(3,1) = 0.00001;
  U1(4,1) = 0.00001;
  U1(5,1) = 0.00001;
  U1(6,1) = 0.00001;

  %IMM - U initialization
  [rows,cols] = size(ekfs);
  n = cols;
  for r=1:n  %total number of KF
     for c=1:n  %total number of KF
        U(r,c) = 0.00001;
     end
  end

  %IMM - Transition Probability Matrix
  BT(1, 1)=0.305825242718447;
  BT(1, 2)=0.635922330097087;
  BT(1, 3)=0.058252427184466;
  BT(2, 1)=0.0411089866156788;
  BT(2, 2)=0.445028680688337;
  BT(2, 3)=0.513862332695985;
  BT(3, 1)=0.0211424332344214;
  BT(3, 2)=0.385385756676558;
  BT(3, 3)=0.593471810089021;

     return;
  end
```

The **ekf_correct** and **ekf_predict** steps, as their names imply, correspond to the mathematical functions used in the correct and predict steps for the KF estimations.

```
function [X2, P2]=ekf_correct(ekf, H, P0, R, I, Z, X0)

   HPHT = H*P0*H';
   S = HPHT + R;
   k1 = P0*H';
   %--------
   %Determine how many elements in diagonal are important
   for this KF
   if ekf == 1
      d = 2;
   elseif ekf == 2
      d = 4;
   else
      d = 6;
   end
   %--removing zeros from the diagonal to be able to do
the inverse
   [rows,cols] = size(S);
   for r=d+1:rows
      for c=d+1:cols
         if r == c
            if S(r,c) > -0.0001 & S(r,c) < 0.0001
               S(r,c) = 1;
            end
         end
      end
   end
   %--doing the inverse of S
   k2 = inv(S);
   %--adding zeros back to the diagonal to maintain
matrix properties
   [rows,cols] = size(k2);
   for r=d+1:rows
      for c=d+1:cols
         if r == c & k2(r,c) == 1
            k2(r,c) = 0;
         end
      end
   end
   %--------
   k = k1*k2;
   X2 = X0+(k*(Z-X0));
   P2 = (I-(k*H)) * P0;
   return;
end
```

```
function [P2]=ekf_predict(status, A, Pf, Qf)

    APAT = A*Pf*A';
    P2   = APAT+Qf;

   return;
end
```

The **ekf_models** function is used by the ekf_prefict function to execute all the different KF models defined to estimate the next position.

```
function [X2]=ekf_models(rec,sensor, ekf, status, Dk,
Dk_orig, Z, Z_prev, gps_dir,acc_avx,acc_avy,time)
%Reference of data rows
  %Z(1,1)= gps_lon;
  %Z(2,1)= gps_lat;
  %Z(3,1)= velx;
  %Z(4,1)= vely;
  %Z(5,1)= acc_avx;  is nominal acc (perpendicular to the
direction of movement; positive means towards the right)
  %Z(6,1)= acc_avy;  is tangential acc (in the direction
of movement; positive means forward)

   X2 = Z;

   %---EKF1 - const_location
   if ekf == 1
      X2(1,1)  = Z(1,1);
      X2(2,1)  = Z(2,1);
      X2(3,1)  = 0;
      X2(4,1)  = 0;
      X2(5,1)  = 0;
      X2(6,1)  = 0;
   end

   %---EKF2 - const_speed
   if ekf == 2
      X2(1,1)  = Z(1,1)  + Z(3,1)*Dk;
      X2(2,1)  = Z(2,1)  + Z(4,1)*Dk;
      X2(3,1)  = Z(3,1); %constant velocity
      X2(4,1)  = Z(4,1); %constant velocity
      X2(5,1)  = 0;      %no acceleration
```

```
        X2(6,1) = 0;          %no acceleration
    end

    %---EKF3 - const_acc
    if ekf == 3
        An=Z(5,1); %Saving An (assume it does not change
through the next 3s)
        At=Z(6,1); %Saving At (assume it does not change
through the next 3s)
        if At>0
            Ax = ( At*Z(3,1) + An*Z(4,1) ) / sqrt( Z(3,1)^2
+ Z(4,1)^2 );
            Ay = ( At*Z(4,1) - An*Z(3,1) ) / sqrt( Z(3,1)^2
+ Z(4,1)^2 );
        else
            Ax = ( At*Z(3,1) - An*Z(4,1) ) / sqrt( Z(3,1)^2
+ Z(4,1)^2 );
            Ay = ( At*Z(4,1) + An*Z(3,1) ) / sqrt( Z(3,1)^2
+ Z(4,1)^2 );
        end
        Vx = Z(3,1) + Ax *Dk;
        Vy = Z(4,1) + Ay *Dk;
        %--- EKF3 models
        X2(1,1) = Z(1,1) + ( Z(3,1) )*Dk + (1/2)*( Ax
)*Dk^2;
        X2(2,1) = Z(2,1) + ( Z(4,1) )*Dk + (1/2)*( Ay
)*Dk^2;
        X2(3,1) = Vx;
        X2(4,1) = Vy;
        X2(5,1) = Z(5,1); %constant acceleration sideways
        X2(6,1) = Z(6,1); %constant acceleration forwards
    end

    return;
end
```

The **ekf_update** is used to update the Q matrix of the KF for the DRWDE research done in Chapter 2 of this book.

```
function [R,Q]=ekf_update(status,ekfs,Dk,Dks1,Dks2,Dks3,
use_Q_calc_vars,use_R_calc_vars)

%default basic values
sig_p = 0.01;
sig_m = 1;
%max measurements
vel_max = abs(37.5513600);
```

```
v          = vel_max;
acc_t_max = abs(-4.9425516);
acc_n_max = abs(-4.3541526);
a          = acc_t_max;
acc_Dt_max = abs(5.8839900);
acc_Dn_max = abs(7.3549875);
j           = acc_Dt_max;

%sensors' built it errors (assumption)
b1=0.1; %GPS
b2=0.1; %GPS
b3=0.1; %ST
b4=0.1; %ST
b5=0.1; %acc
b6=0.1; %acc

%Loop through all EKFs to set their specific Q arrays
 [rows,cols] = size(ekfs);
n = cols;
for f=1:n

   %---- KF - Q
   if use_Q_calc_vars == 1
      if f == 1
Q2(1,1)=(1/36)*(j^2)*(Dk^6); Q2(1,2)=0;        Q2(1,3)=0;
Q2(1,4)=0;     Q2(1,5)=0;        Q2(1,6)=0;
Q2(2,1)=0;     Q2(2,2)=(1/36)*(j^2)*(Dk^6);    Q2(2,3)=0;
Q2(2,4)=0;     Q2(2,5)=0;        Q2(2,6)=0;
Q2(3,1)=0;     Q2(3,2)=0;        Q2(3,3)=0;
Q2(3,4)=0;     Q2(3,5)=0;        Q2(3,6)=0;
Q2(4,1)=0;     Q2(4,2)=0;        Q2(4,3)=0;
Q2(4,4)=0;     Q2(4,5)=0;        Q2(4,6)=0;
Q2(5,1)=0;     Q2(5,2)=0;        Q2(5,3)=0;
Q2(5,4)=0;     Q2(5,5)=0;        Q2(5,6)=0;
Q2(6,1)=0;     Q2(6,2)=0;        Q2(6,3)=0;
Q2(6,4)=0;     Q2(6,5)=0;        Q2(6,6)=0;
        Q{f}=Q2;
      elseif f == 2
Q2(1,1)=(1/36)*(j^2)*(Dk^6);     Q2(1,2)=0;
Q2(1,3)=(1/12)*(j^2)*(Dk^5);     Q2(1,4)=0;
Q2(1,5)=0;  Q2(1,6)=0;
Q2(2,1)=0;  Q2(2,2)=(1/36)*(j^2)*(Dk^6);     Q2(2,3)=0;
Q2(2,4)=(1/12)*(j^2)*(Dk^5);     Q2(2,5)=0;  Q2(2,6)=0;
Q2(3,1)=(1/12)*(j^2)*(Dk^5);     Q2(3,2)=0;
Q2(3,3)=(1/4)*(j^2)*(Dk^4);      Q2(3,4)=0;
Q2(3,5)=0;      Q2(3,6)=0;
Q2(4,1)=0;             Q2(4,2)=(1/12)*(j^2)*(Dk^5);
Q2(4,3)=0;             Q2(4,4)=(1/4)*(j^2)*(Dk^4);
Q2(4,5)=0;             Q2(4,6)=0;
```

```
Q2(5,1)=0;                  Q2(5,2)=0;
Q2(5,3)=0;                  Q2(5,4)=0;                  Q2(5,5)=0;
Q2(5,6)=0;
Q2(6,1)=0;                  Q2(6,2)=0;                  Q2(6,3)=0;
Q2(6,4)=0;                  Q2(6,5)=0;                  Q2(6,6)=0;
         Q{f}=Q2;
      elseif f == 3
Q2(1,1)=(1/36)*(j^2)*(Dk^6);      Q2(1,2)=0;
Q2(1,3)=(1/12)*(j^2)*(Dk^5);      Q2(1,4)=0;
Q2(1,5)=0;                        Q2(1,6)=0;
Q2(2,1)=0;                  Q2(2,2)=(1/36)*(j^2)*(Dk^6);
Q2(2,3)=0;                  Q2(2,4)=(1/12)*(j^2)*(Dk^5);
Q2(2,5)=0;                  Q2(2,6)=0;
Q2(3,1)=(1/12)*(j^2)*(Dk^5);      Q2(3,2)=0;
Q2(3,3)=(1/4)*(j^2)*(Dk^4);       Q2(3,4)=0;
Q2(3,5)=0;                  Q2(3,6)=0;
Q2(4,1)=0;                  Q2(4,2)=(1/12)*(j^2)*(Dk^5);
Q2(4,3)=0;                  Q2(4,4)=(1/4)*(j^2)*(Dk^4);
Q2(4,5)=0;                  Q2(4,6)=0;
Q2(5,1)=0;                  Q2(5,2)=0;
Q2(5,3)=0;                  Q2(5,4)=0;
Q2(5,5)=(j^2)*(Dk^2);       Q2(5,6)=0;
Q2(6,1)=0;                  Q2(6,2)=0;
Q2(6,3)=0;                  Q2(6,4)=0;
Q2(6,5)=0;                  Q2(6,6)=(j^2)*(Dk^2);
Q{f}=Q2;
      end
   else
      for z1=1:6
         for z2=1:6
            Q{f}(z1,z2) = 0;
         end
         Q{f}(z1,z1) = sig_p^2;
      end
   end

   %---- KF - R
   for z1=1:6
      for z2=1:6
         R{f}(z1,z2) = 0;
      end
      R{f}(z1,z1) = sig_p^2;
   end

end

   return;
end
```

The **imm_part1** and **imm_part2** functions are used, as their names imply, to execute the mathematical equations for the IMM parts of the trajectory estimation framework.

```
function [X0,P0,cb0,U0]=imm_part1(ekf_status, sensors,
ekfs, X, P, BT, U1, U, cb)
%U0 is passed back only to have it available when passed
back into this function as "U", it is not used outside of
this function

%Get total number of EKFs defined
[rows,cols] = size(ekfs);
total_ekfs=cols;

%--- IMM step 1 --- Calculation of the mixing
probabilities
for c=1:total_ekfs
    cb0(c,1)=0;
    for r=1:total_ekfs
        ttt=cb0(c,1);
        cb0(c,1)=cb0(c,1)+BT(r,c)*U1(r,1);
    end
    if(cb0(c,1) <= 0)
        cb0(c,1)=0.0001;
    end
    for r=1:total_ekfs
            U0(r,c)=(1/cb0(c,1))*BT(r,c)*U1(r,1);
        if(U0(r,c) <= 0)
            U0(r,c)=0.0001;
        end
    end
end

%--- IMM step 2 --- Mixing
for f=1:total_ekfs
    X_ekf  = X{f};
    for r=1:6   %total rows in Z
        X0_ekf(r,1)=0;
        for c=1:total_ekfs
            X0_ekf(r,1)= X0_ekf(r,1)+ (X_ekf(r,1)*U0(c,f));
        end
        errj0_(r,1)= X_ekf(r,1)-X0_ekf(r,1);
    end
    errj0{f} = errj0_*errj0_';
    X0{f} = X0_ekf;
end

for f=1:total_ekfs
```

```
   for r=1:6   %total rows in Z
      for c=1:6   %total rows in Z
         P0_ekf(r,c)=0;
         for j=1:total_ekfs
            P_ekf = P{j};
            errj0_ekf = errj0{j};
            P0_ekf(r,c) = P0_ekf(r,c) +
(P_ekf(r,c)+errj0_ekf(r,c))*U0(j,f);
         end
      end
   end
   P0{f} = P0_ekf;
end

   return;
end



function [U1,X_imm,mm_f1]=imm_part2(ekf_status, sensors,
ekfs, X, P, H, R, cb, mm_f)

%Get total number of KFs in use
[rows,cols] = size(ekfs);
total_ekfs=cols;

%--- IMM step 3 --- Mode matched filtering
for f=1:total_ekfs
   X_ekf = X{f};
   P_ekf = P{f};
   R_ekf = R{f};

   mm_s2   = abs(det(H*P_ekf*H'+R_ekf));
   HPHT = H*P_ekf*H';

   %---Determine how many elements in diagonal are
important for this KF
   if ekfs == 1
      d = 2;
   elseif ekfs == 2
      d = 4;
   else
      d = 4; %d=6 does not seem to be working
   end
   %--removing zeros from the digonal to be able to do
the inverse
   [rows,cols] = size(HPHT);
```

```
    for r=d+1:rows
       for c=d+1:cols
          if r == c
             if HPHT(r,c) > -0.000001 & HPHT(r,c) <
0.000001
                HPHT(r,c) = 1;
             end
          end
       end
    end
    %--doing the inverse of HPHT
    HPHT_inv = inv(HPHT);
    %--adding zeros back to the diagonal to maintain
matrix properties
    [rows,cols] = size(HPHT_inv);
    for r=d+1:rows
       for c=d+1:cols
          if r == c & HPHT_inv(r,c) == 1
             HPHT_inv(r,c) = 0;
          end
       end
    end
    %--------
    mm_x2    = det( HPHT_inv ) ;
    mm_f_ekf = (1/sqrt( ((2*pi)^(total_ekfs/2) )*mm_s2 ))
*exp((-1/2)*mm_x2);

    if(mm_f_ekf <= 0)
       mm_f_ekf=0.0001;
    end

    mm_f1(f,1) = mm_f_ekf;

end

%--- IMM step 4 --- Mode probability update
c = 0;
for f=1:total_ekfs
    ttt = c;
    c = c + mm_f1(f,1)*cb(f,1);
end
if(c < = 0)
    c=0.0001;
end
for f=1:total_ekfs
    U1(f,1) = (1/c) * mm_f1(f,1) * cb(f,1);
    if(U1(f,1) < = 0)
```

```
        U1(f,1)=0.0001;
    end
end

%--- IMM step 5 --- For OUTPUT purposes only (not part of
the algorithm recursions)
for r=1:6   %IMPORTANT: change this number to the total
rows in Z.
   X_imm(r,1) = 0;
   for f=1:total_ekfs
      X_ekf = X{f};
      ttt = X_imm(r,1);
      X_imm(r,1) = X_imm(r,1) + (X_ekf(r,1)*U1(f,1));
   end
end

   return;
end
```

The **correction** function is used when a correction method is enabled, such as use_geom_method=1. It is used to do some type of rounding method to remove spikes in the values, which are probably outliers. It was set up during the research step of this experiment, but was not enabled for the results reported in Chapters 2 and 3.

```
function [Z2,Ap,Cp]=correction(ss0,est,est0,Ap,Cp,ang1,an
g2,Z,Z0,rec,cur_hr,cur_min,cur_sec)

   if ss0 == 0
      %The following variables are used in the use_geom_
method (if enabled)
      if( est(3,1) < 0 || est(3,1) > 0 )
         Ap = est0(4,1)/est0(3,1); %Vy/Vx
      else
         if( ang1 > 0 )
            Ap = -128;
         else
            Ap = 127;
         end
      end
      Cp = est0(2,1) - est0(1,1)*Ap;
   end

   %correction funct
   arriba = 0;
```

```
   aux = ang1-ang2;
   if( aux<2 && aux>-2 )
      if( (Z(1,1)>Z0(1,1) && aux<0) || (Z(1,1)<=Z0(1,1)
&& aux>=0) )
         arriba = 1;
      end
   elseif( ang1<0 )
      arriba = 1;
   end

         Z2 = est;

         if( (arriba == 1 && (est(2,1) <
Ap*est0(1,1)+Cp)) || (arriba == 0 && (est(2,1) >=
Ap*est0(1,1)+Cp)) )
            aux = ( (est0(1,1)-est(1,1))*Ap + est(2,1)
- est0(2,1) ) / ((Ap^2)+1);
            Z2(1,1) = est(1,1) + Ap*aux;
            Z2(2,1) = est(2,1) - aux;

            dist1 = 1000000;
            a_n_final = est(5,1);
            for i=-8:8
               a_n = 0.15*i + est(5,1);
               %Determine new normal acceleration based
on what new location is supposed to be (after correction)
               est2=Z;
               for d=1:30
                  Z3=est2;
                  An=a_n; %Saving An as we will assume it
does not change through the next 3 seconds
                  At=Z3(6,1); %Saving At as we will
assume it does not change through the next 3 seconds
                  if At>0
                     Ax = ( At*Z3(3,1) + An*Z3(4,1) ) /
sqrt( Z3(3,1)^2 + Z3(4,1)^2 );
                     Ay = ( At*Z3(4,1) - An*Z3(3,1) ) /
sqrt( Z3(3,1)^2 + Z3(4,1)^2 );
                  else
                     Ax = ( At*Z3(3,1) - An*Z3(4,1) ) /
sqrt( Z3(3,1)^2 + Z3(4,1)^2 );
                     Ay = ( At*Z3(4,1) + An*Z3(3,1) ) /
sqrt( Z3(3,1)^2 + Z3(4,1)^2 );
                  end
                  Vx = Z3(3,1) + Ax *0.1;
                  Vy = Z3(4,1) + Ay *0.1;
                  est2(1,1) = Z3(1,1) + ( Z3(3,1) )*0.1 +
(1/2)*( Ax )*0.1^2;
```

```
                    est2(2,1) = Z3(2,1) + ( Z3(4,1) )*0.1 +
(1/2)*( Ay )*0.1^2;
                    est2(3,1) = Vx;
                    est2(4,1) = Vy;
                    est2(5,1) = Z3(5,1); %constant
acceleration sideways (normal)
                    est2(6,1) = Z3(6,1); %contant
acceleration forwards (tangential)
                 end
                 dist2 = (Z2(1,1)-est2(1,1))^2 +
(Z2(2,1)-est2(2,1))^2;
                 if dist1 > dist2
                    dist1 = dist2;
                    a_n_final = a_n;
                 end
              end
              Z2(5,1) = a_n_final;

           end

     return;
  end
```

The **format_results2** function is used to properly format the estimated positions into a matrix and to calculate the position errors between the real GPS measurement and the estimated future position.

```
function
[OUT,start_time]=format_results2(data,sensor_
status,sensors,ekfs,use_imm,OUT,Z,X_ahead,X_imm_
ahead,Dk2,Dk,Dk_orig,loop_count,loop_start,rec,est_sec_
ahead,start_time)

   %take time from first record
   cur_hr  = data(rec,1);
   cur_min = data(rec,2);
   cur_sec = data(rec,3);
   cur_time = ( ( ( (cur_hr*60)+cur_min )*60)+cur_sec
)*10;  %multiple x10 to make fractions of the second
integers

   %Record start_time so we can calculate row_count
starting from 0
   if loop_count == loop_start
      start_time = cur_time-1;
```

```
   end

   %Get row number from difference in time
   row_count = cur_time - start_time;
   %Determine time gap in estimation for future location
either Dk or est_sec_ahead
   if est_sec_ahead > 0
      step = round(Dk2*10);    %Dk ahead     (multiply by
10 to get only integer numbers)
   else
      step = round(Dk*10);  %Dk ahead        (multiply by
10 to get only integer numbers)
   end

   %Get total number of EKFs defined
   [rows,cols] = size(ekfs);
   total_ekfs=cols;

        %Recording results
        OUT(row_count,1)=rec;                       %record
record number in first column
        OUT(row_count,2)=cur_hr;
        OUT(row_count,3)=cur_min;
        OUT(row_count,4)=cur_sec;
        OUT(row_count,5)=Dk;

        OUT(row_count,7)=sensor_status(1,1);  %record
status of sensor 1
        OUT(row_count,8)=sensor_status(2,1);  %record
status of sensor 2
        OUT(row_count,9)=sensor_status(3,1);  %record
status of sensor 3

        OUT(row_count,11)=step;                    %record step

        n=13; %start from this column number
        OUT(row_count,n)=Z(1,1);                   %GPS
latitude on first column of this group
        for c=1:total_ekfs  %loop through defined EKFs
           OUT(row_count+step,n+c)=X_ahead{c}(1,1);
        end
        if use_imm == 1  %record IMM result on last
column of this group
           OUT(row_count+step,n+1+c)=X_imm_ahead(1,1);
        end

        n=n+1+c+2; %leave one column empty in between lat
and lon sections (the column will have all zeros)
```

```
        OUT(row_count,n)=Z(2,1);                   %GPS
longitude on first column of this group
        for c=1:total_ekfs  %loop through defined EKFs
           OUT(row_count+step,n+c)=X_ahead{c}(2,1);
        end
        if use_imm == 1  %record IMM result on last
column of this group
           OUT(row_count+step,n+1+c)=X_imm_ahead(2,1);
        end

        n=n+1+c+2; %leave one column empty in between lat
and lon sections (the column will have all zeros)
        OUT(row_count,n)=Z(3,1);                   %Z  heading
on first column of this group
        for c=1:total_ekfs  %loop through defined EKFs
           OUT(row_count+step,n+c)=X_ahead{c}(3,1);
        end
        if use_imm == 1  %record IMM result on last
column of this group
           OUT(row_count+step,n+1+c)=X_imm_ahead(3,1);
        end

        n=n+1+c+2; %leave one column empty in between lat
and lon sections (the column will have all zeros)
        OUT(row_count,n)=Z(4,1);                   %Z  speed on
first column of this group
        for c=1:total_ekfs  %loop through defined EKFs
           OUT(row_count+step,n+c)=X_ahead{c}(4,1);
        end
        if use_imm == 1  %record IMM result on last
column of this group
           OUT(row_count+step,n+1+c)=X_imm_ahead(4,1);
        end

        n=n+1+c+2; %leave one column empty in between lat
and lon sections (the column will have all zeros)
        OUT(row_count,n)=Z(5,1);                   %Z  acc_x on
first column of this group
        for c=1:total_ekfs  %loop through defined EKFs
           OUT(row_count+step,n+c)=X_ahead{c}(5,1);
        end
        if use_imm == 1  %record IMM result on last
column of this group
           OUT(row_count+step,n+1+c)=X_imm_ahead(5,1);
        end
```

```
        n=n+1+c+2; %leave one column empty in between lat
and lon sections (the column will have all zeros)
        OUT(row_count,n)=Z(6,1);            %Z  acc_y on
first column of this group
        for c=1:total_ekfs  %loop through defined EKFs
           OUT(row_count+step,n+c)=X_ahead{c}(6,1);
        end
        if use_imm == 1  %record IMM result on last
column of this group
           OUT(row_count+step,n+1+c)=X_imm_ahead(6,1);
        end

   return;
end
```

Index